T0142849

Lecture Notes in Electrical Engineering

Volume 713

The book series *Lecture Notes in Electrical Engineering* (LNEE) publishes the latest developments in Electrical Engineering - quickly, informally and in high quality. While original research reported in proceedings and monographs has traditionally formed the core of LNEE, we also encourage authors to submit books devoted to supporting student education and professional training in the various fields and applications areas of electrical engineering. The series cover classical and emerging topics concerning:

- Communication Engineering, Information Theory and Networks
- Electronics Engineering and Microelectronics
- Signal, Image and Speech Processing
- Wireless and Mobile Communication
- Circuits and Systems
- Energy Systems, Power Electronics and Electrical Machines
- Electro-optical Engineering
- Instrumentation Engineering
- Avionics Engineering
- Control Systems
- Internet-of-Things and Cybersecurity
- Biomedical Devices, MEMS and NEMS

For general information about this book series, comments or suggestions, please contact leontina.dicecco@springer.com.

To submit a proposal or request further information, please contact the Publishing Editor in your country:

China

Jasmine Dou, Associate Editor (jasmine.dou@springer.com)

India, Japan, Rest of Asia

Swati Meherishi, Executive Editor (Swati.Meherishi@springer.com)

Southeast Asia, Australia, New Zealand

Ramesh Nath Premnath, Editor (ramesh.premnath@springernature.com)

USA, Canada:

Michael Luby, Senior Editor (michael.luby@springer.com)

All other Countries:

Leontina Di Cecco, Senior Editor (leontina.dicecco@springer.com)

**** Indexing: Indexed by Scopus. ****

More information about this series at http://www.springer.com/series/7818

Marcin Kubica · Adam Opara ·
Dariusz Kania

Technology Mapping for LUT-Based FPGA

Marcin Kubica
Silesian University of Technology
Gliwice, Poland

Adam Opara
Silesian University of Technology
Gliwice, Poland

Dariusz Kania
Silesian University of Technology
Gliwice, Poland

ISSN 1876-1100 ISSN 1876-1119 (electronic)
Lecture Notes in Electrical Engineering
ISBN 978-3-030-60490-5 ISBN 978-3-030-60488-2 (eBook)
https://doi.org/10.1007/978-3-030-60488-2

This Springer imprint is published by the registered company Springer Nature Switzerland AG
The registered company address is: Gewerbestrasse 11, 6330 Cham, Switzerland

Preface

This book is a summary of the authors' many years of research in the field of logic synthesis. These studies concentrated in the areas of function decomposition, technological mapping or recently cyber-physical systems. These issues were the subject of two doctoral dissertations: Adam Opara: "*Dekompozycyjne metody syntezy układów kombinacyjnych wykorzystujące binarne diagramy decyzyjne*" (2009) and Marcin Kubica: "*Dekompozycja i odwzorowanie technologiczne z wykorzystaniem binarnych diagramów decyzyjnych*" (2014) in Silesian University of Technology. The ideas underlying these Ph.D. theses have been developed over the years, which has been the basis of many scientific publications that are referenced in the book.

The motivation for the book was the desire to present comprehensively the issues that the authors have dealt with in recent years. The book is a presentation of a number of ideas published in earlier articles, to which the reader is directed and the appropriate reference materials. This type of approach allows a more complete presentation of the essence of problems and solutions proposed by the authors, meticulously presented in the indicated articles. The authors emphasize that this book is a review, and the original source of ideas presented are the authors' publications and doctoral dissertations, from which the authors took materials necessary to create this book.

We would like to thank the reviewers and colleagues for many valuable comments, which were the inspiration for continuous improvement of developed methods and influenced the final form of this book. We thank wives, daughters and all relatives for invaluable help and understanding.

Gliwice, Poland

Marcin Kubica
Adam Opara
Dariusz Kania

Preface

Contents

1 Introduction . . . 1
- 1.1 ASIC Implementation of Digital Circuits . . . 1
- 1.2 Influence of Architecture on the Way of Conducting Logic Synthesis . . . 5
- 1.3 Academic Systems of Logic Synthesis Oriented at FPGA . . . 6
- 1.4 System on Chip . . . 8
- References . . . 9

2 Methods for Representing Boolean Functions—Basic Definitions . . . 15
- 2.1 Hypercube, Cube, Implicant, Minterm . . . 15
- 2.2 Two-Level Description of a Boolean Function . . . 18
- 2.3 Truth Table, Karnaugh Map, Binary Decision Tree . . . 20
- References . . . 23

3 Binary Decision Diagrams . . . 25
- 3.1 BDD Operations . . . 29
- 3.2 Basics of Software Implementation . . . 31
  - 3.2.1 Representation of Diagram Nodes in Computer Memory . . . 31
  - 3.2.2 Negation Attribute . . . 33
- References . . . 35

4 Theoretical Basis of Decomposition . . . 39
- 4.1 Functional Decomposition Theorem . . . 39
- 4.2 Complex Decomposition Models . . . 41
- 4.3 Iterative Decomposition . . . 42
- 4.4 Multiple Decomposition . . . 45
- 4.5 Direction of Decomposition . . . 48
- References . . . 49

5 Decomposition of Functions Described Using BDD 53
5.1 Methods for Performing Function Decomposition Using Single Cuts of the BDD 53
5.1.1 Simple Serial Decomposition—Single Cut Method 53
5.1.2 Iterative Decomposition—Single Cut Method 55
5.1.3 Multiple Decomposition—A Single Cut Method 56
5.2 Methods of Realizing Function Decomposition Using Multiple Cuts of the BDD Diagram 57
5.2.1 SMTBDD in the Decomposition Process 59
5.2.2 Simple Serial Decomposition—Multiple Cutting Method 60
5.2.3 Multiple Decomposition—Multiple Cutting Method 60
References 63

6 Ordering Variables in BDD Diagrams 65
References 69

7 Nondisjoint Decomposition 71
References 75

8 Decomposition of Multioutput Functions Described Using BDD 77
8.1 Creating Common Bound Blocks 77
8.2 Method of Creating Multioutput Function 82
8.3 Methods of Merging Single Functions into Multioutput Using PMTBDD 86
References 91

9 Partial Sharing of Logic Resources 93
9.1 Equivalence Classes 96
9.2 Partial Sharing in SMTBDD 97
9.3 Searching for Equivalence Classes with MTBDD Usage 98
References 112

10 Ability of the Configuration of Configurable Logic Blocks 115
10.1 Configuration Features of Logic Cells 115
References 117

11 Technology Mapping of Logic Functions in LUT Blocks 119
11.1 Selection of Cutting Line 119
11.2 Methods for Determining the Efficiency of Technology Mapping 122
11.3 Triangle Tables, Including Nondisjoint Decomposition 126
11.4 Technology Mapping that Takes into Account the Sharing of Logic Resources 128
References 131

12 Technology Mapping of Logic Functions in Complex Logic Blocks . . . 133
12.1 Technology Mapping in ALM Blocks . . . 133
12.2 Methods for Assessing Technology Mapping in ALM Blocks . . . 140
12.3 Technology Mapping of Nondisjoint Decomposition . . . 144
References . . . 145

13 Decomposition Methods of FSM Implementation . . . 147
13.1 Finite State Machine Background . . . 147
13.2 Technology Mapping of the FSM Combination Part in Configuration Logic Blocks . . . 148
References . . . 150

14 Algorithms for Decomposition and Technological Mapping . . . 153
References . . . 156

15 Results of Experiments . . . 159
15.1 Comparison of MultiDec and DekBDD Systems . . . 159
15.2 Comparison with Selected Academic Systems . . . 163
15.3 Effect of Triangle Tables on the Results . . . 169
15.4 Comparison of the MultiDec System with Commercial Systems . . . 189
15.5 Synthesis of Sequential Circuits . . . 189
15.6 Technology Mapping in Complex Logic Blocks—Results of Experiments . . . 191
References . . . 199

16 Summary . . . 201

Index . . . 205

Chapter 1
Introduction

The design process of digital systems requires the use of specialized computer-aided design (CAD) software. The diversity of digital circuit implementations and the domination of application-specific integrated circuits (ASICs) create many problems in the field of automatic synthesis. Describing the designed systems and converting description to a form implemented in hardware are challenging. These problems have contributed to the development of high-level forms of system description, which may include hardware description languages or even system description languages.

The description of a system at a high level of abstraction, for example, using hardware description languages (VHDL, VERILOG HDL), renders the process of compiling various language forms extremely important. A very important issue of modern logic synthesis is linking the elements of synthesis with the problems of technology mapping of circuits. Only effective synthesis algorithms, including technology mapping methods that are well suited to the architecture of circuits, can enable "good" system design (effectively using the logic resources of the structure).

The intention of the authors of this book is to present various aspects of modern logic synthesis with a focus on the issues of technology mapping of designed circuits in Field-Programmable Gate Array (FPGA) structures. The issues of synthesis and technology mapping are discussed primarily in connection with binary decision diagrams.

1.1 ASIC Implementation of Digital Circuits

In the group of currently produced integrated digital circuits, a person can distinguish standard circuits, which are intended for general, universal applications and specialized circuits (Application-Specific Integrated Circuits - ASIC) that are designed or adapted to the individual needs of the user.

Implementations of digital systems based on standard integrated circuits that are selected from an extensive range of mass-produced components currently do not have

M. Kubica et al., *Technology Mapping for LUT-Based FPGA*, Lecture Notes in Electrical Engineering 713, https://doi.org/10.1007/978-3-030-60488-2_1

economic justification, even with a small production series. Solutions in the form of various forms of ASICs are substantially more profitable. Integrated circuits of this type are manufactured or adapted to the individual needs of the customer. Several types of circuits can be distinguished in the ASIC circuit group. They differ in the degree of resource individualization, method of initial preparation of gate matrices and method of adapting the "semifinished products" to the needs of the recipient.

Four groups of ASICs exist: full custom, standard cells, gate arrays and programmable logic devices. In the case of full custom circuits, an entire system is created on a "clean" silicon surface, which renders it profitable only for the largest production series. Designing circuits in the form of gate arrays and standard cells requires the use of prepared library elements that are adapted to the applied technology or semifinished products that contain appropriate matrices of elementary elements (transistors, and resistors). In each case, an appropriate specification of the digital system is combined with the selection of library elements. The process of creating an integrated circuit completes the execution of masks that enable you to connect the components.

For small production runs, PLDs are utilized (PLD—Programmable Logic Devices). In this case, the process of adapting a circuit to the user's needs is achieved by programming. The development of programmable devices has enabled them to address a growing area of applications. The key advantage of this form of implementation of an integrated circuit is its easy modification, which consists of reprogramming a device. This type of feature predisposes programmable structures to applications that are aimed at prototyping. In addition, this type of solution does not require the use of very advanced technology lines, and the entire process of manufacturing a digital system can occur at the designer's desk. The constantly increasing complexity of programmable structures and the increasing availability of software affect the high attractiveness of this type of solution. A significant advantage of this technology is the possibility of reprogramming a system in a manufactured and operating digital system.

Field Programmable Gate Arrays are currently the market-dominating family of PLDs. They consist of programmable logic cells (blocks) and additional logic resources for connecting them and exchanging information with the environment. The most popular FPGAs are LUT-based FPGAs [1, 2] that contain configurable logic blocks, I/O blocks (IOB) and interconnect areas that are used to make connections. Configurable logic blocks enable the implementation of each logic function with a specific and usually small number of variables. The logic structure of these blocks is similar to the structure of the first programmable read only memory (PROM) types. Configuration data, which are responsible for the way connections are made inside a device, are most often contained in RAM memory cells. The simplified structure of FPGAs is shown in Fig. 1.1. A configurable logic block usually contains LUT array blocks with few inputs but enable any function. A LUT-type logic block that contains n-inputs and m-outputs is abbreviated LUTn/m.

In specific solutions, larger structures are created, e.g., referred to as "slice" cells in Xilinx devices, and in addition to LUT4/1 blocks, they contain additional programmable multiplexers, programmable connections, and elementary arithmetic

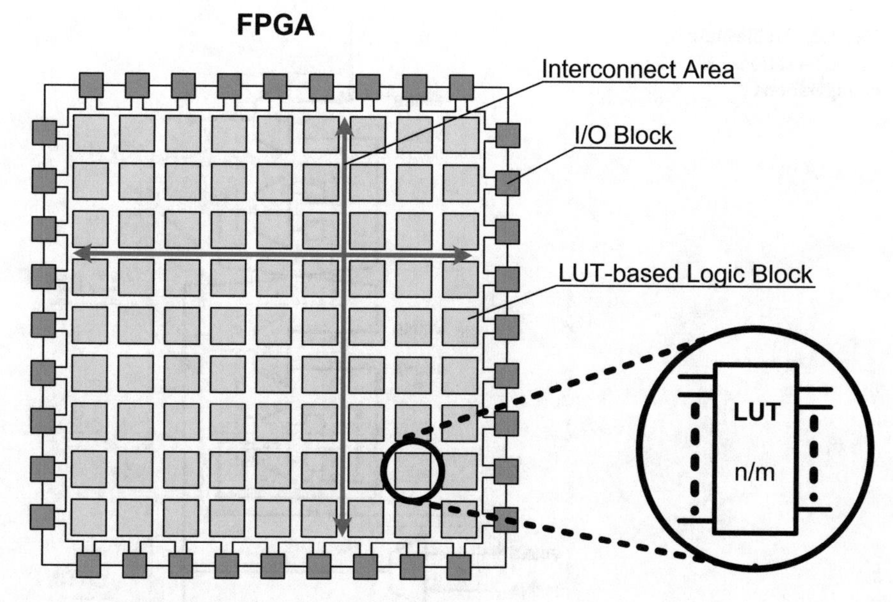

Fig. 1.1 Architecture of LUT-based FPGAs

blocks. The basic element of these solutions is the LUT block, which works in various configurations. Sometimes larger structures can be created within a block; however, the resources of neighboring cells are always utilized, creating significant implementation restrictions [2]. An example of logic resources available in slice blocks of Spartan 3 devices enable the creation of LUT blocks with more inputs than the elementary LUT5/1 block, as shown in Fig. 1.2.

Note that arithmetic elements appear increasingly often in logic blocks of FPGAs or outside of them. These elements include fast transfer circuits or specialized multipliers. FPGA structures are used to accelerate calculations by hardware implementation, which explains the emphasis on optimizing arithmetic operations. A consequence of this is embedding in the FPGA structure numerous dedicated modules, enabling, for example, rapid multiplication and increasing the efficiency of various types of computational applications.

A characteristic feature of FPGA structures is the segmental nature of the connections (Fig. 1.3). Each connection point can be modeled as an RC system with delay. The propagation time introduced by the connection paths depends on the location of the connected blocks inside the FPGA. In addition, the connection resources in FPGAs are limited, making the placement of blocks within the structure (placement, fitting) and the way of routing extremely important. In addition to general purpose connections, the FPGA has a limited number of long lines—usually vertical and horizontal—which create less delay and are usually employed to distribute time-critical signals.

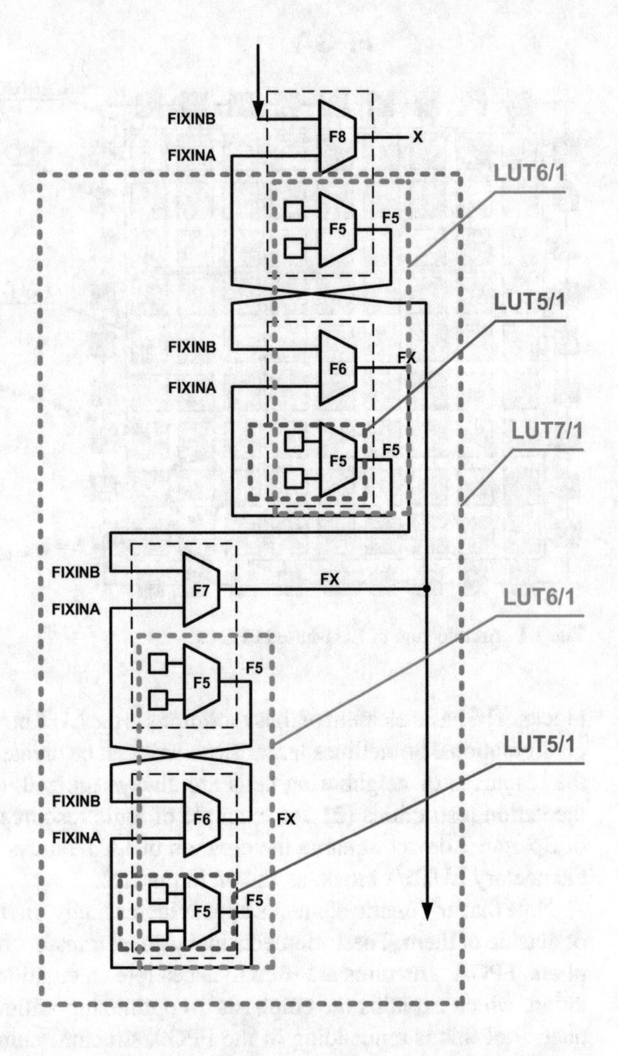

Fig. 1.2 Architecture of "Slice"—various LUT block configurations

Other structures of elementary blocks occur in less popular devices. Actual devices, which are sometimes referred to as multiplexer-based FPGAs, are good example. In this case, the basic element of the configurable block is the multiplexer that operates in various configurations [3]. In addition to FPGAs, CPLDs that consist of matrix of AND gates variously connected to configurable logic cells maintain a significant place in the area of programmable systems. Although the area of application of CPLD devices has significantly decreased in recent years, these device have an important role, especially in energy-saving systems. New synthesis algorithms that focus on the effective use of logic resources of PAL-based CPLDs have been developed [4–10].

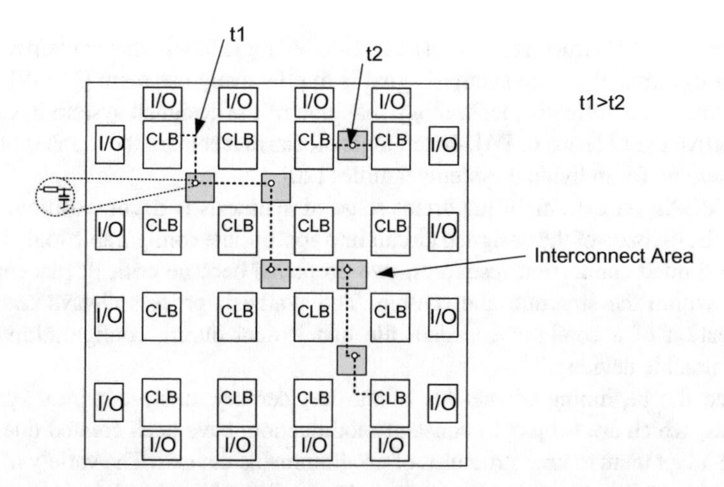

Fig. 1.3 Programmable connections in the FPGA structure

1.2 Influence of Architecture on the Way of Conducting Logic Synthesis

The specificity of programmable devices makes the synthesis process different from the classic synthesis of digital circuits that are implemented using separate elements, i.e., gates, flip-flops, or larger functional blocks, such as multiplexers and counters. The synthesis process in programmable structures is reduced to "fitting" the designed circuit into the programmable structure. The designer's task is to properly describe the circuit and verify the results. The synthesis process is usually automatically performed. Unfortunately, the results are not optimal [11–14].

In the first design stage, the task of the designer is to specify the designed circuit. Various HDL languages are commonly employed for this purpose. A traditional circuit description is possible using a schematic editor and library elements that were created and optimized for the selected programmable structure.

The next stage is the compilation of the description, which is usually integrated with minimization procedures. The large variety and complexity of programmable structures means that one of the basic stages of synthesis is to divide a project into appropriate parts, which are implemented in individual PLDs or logic blocks that occur in CPLDs and FPGAs. Simple programmable devices (PLA, FPLA) did not introduce any restrictions related to the possibilities of connections within a structure. A similar situation applies to CPLDs. The most important stages of the synthesis in this case are two closely related processes: the process of dividing a project into appropriate blocks and the selection of appropriate programmable structures (logic blocks) for their implementation. In the case of simple programmable logic devices (PAL, PLA), the problem is searching for implementation obtained as a result of a division of modules on separate circuits [15]. In the case of CPLDs, the division involves the division of the project into individual logic blocks within a structure. In most cases, CPLD structures consist of PAL-based logic blocks that contain a certain number of terms; thus, the synthesis goal is to effectively use them [16–19]. In the background of an effective technology mapping of the designed system in CPLDs, an effective use of terms or PAL-based logic blocks that contain the number of terms characteristic for individual systems families [20].

In FPGAs, an extremely important stage of synthesis is decomposition, which causes the division of the designed circuit into appropriate configurable logic blocks. Due to limited connection resources, two problems become critical: placement of blocks within the structure and routing. The synthesis process always ends with the creation of a configuration data file that is sent during configuration to the programmable device.

Since the beginning of the 90s of the last century, many complex synthesis methods, which are subject to constant modification, have been created due to the need to adapt them to new structures of programmable devices. The variety of forms of digital circuit implementation enables the creation of universal design tools that are dedicated to the entire range of PLDs from various manufacturers. The advantage of these design support systems is their versatility and the associated ability to transfer projects between different families of programmable devices. The final stages of technology mapping are performed in a company's tools, which are developed by a specific manufacturer of PLDs. Because the target structure is not included in the initial stage of synthesis, universal tools usually do not ensure the efficient use of programmable device resources, although recently this rule does not apply. Unfortunately, portability usually does not correlate with the resource efficiency of programmable structures.

The second group of design support tools are specialized tools included in the company's software packages provided by manufacturers of PLDs. These tools usually enable a more efficient use of specific properties of structures but can be used for a limited group of programmable devices, which is a serious disadvantage. A simplified block diagram of the design process is shown in Fig. 1.4.

1.3 Academic Systems of Logic Synthesis Oriented at FPGA

Logic synthesis is the subject of many scientific papers. As a rule, the work is limited to a single stage of synthesis. In the 1980s, the subject of function minimization was popular and produced algorithms such as Espresso [21]. In the 90s and the beginning of the twenty-first century, work focused on topics related to function decomposition and technologic mapping.

The first logic synthesis algorithms dedicated to an LUT-based FPGA were based on logic synthesis collaborated to gate-based circuits (MIS-PGA [22, 23], ASYL [24, 25], Chortle [26]). These algorithms used various elements of multilevel optimization, factorization of Boolean expressions, variable partitioning, lexicographical

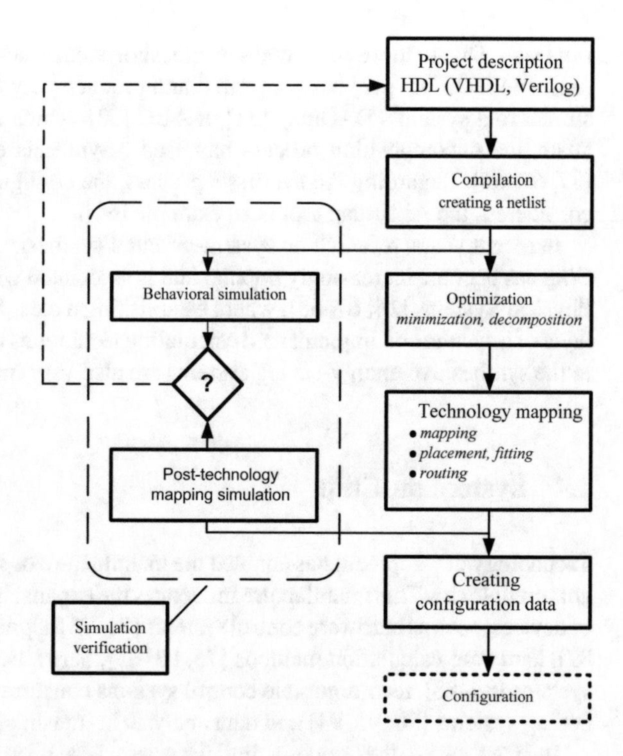

Fig. 1.4 Block diagram of the design process

ordering of variables, and dynamic programming. LUT mapping steps were often based on the iterative division of a network of gates. In the initial step, the network of gates corresponded to a Boolean expression after a two-level minimization. Network mapping was based on an appropriate choice of nodes [26–30]. The technology mapping process transforms a technology-independent logic network into a network of logic nodes. Each logic node was represented by a k-input LUT. State-of-art FPGA technology mappers, which are based on a division of a logic network, are presented in [31–33].

Decomposition is a crucial element of a logic synthesis dedicated to an LUT-based FPGA. Decomposition enables a project to be divided into the parts that have a given number of inputs, because LUTs enable a function that has a limited number of variables (4, 5 or 6) to be carried out. Ashenhurst-Curtis theory [25, 34, 35] is a background for many decomposition algorithms that are oriented to LUT-based FPGAs. This model of decomposition became the basis of the second group of technology mapping approaches [26, 36–40]. These algorithms often use graph coloring techniques [40–44], function transformation [40, 45], and a combination of classic decomposition with the procedures of outputs' division. A classic decomposition model is used in the synthesis process that is oriented to PAL-based CPLD [7, 46, 47]. The main interest, however, is concerned with the methods that use Binary Decision Diagrams (BDD) [48–59]. Description of logic functions in the form of BDD may cause substantial limitation of memory usage and makes the operations run faster. Due to these advantages, synthesis systems such as BDS [60], BDS–PGA [61], and DDBDD [62] have appeared and produce very efficient solutions that are similar to a system's DAOmap [31] or ABC [63], which use AIG graphs. In recent years, the decomposition process has used resynthesis elements more frequently [37, 64, 65]. Regarding the synthesis process, the configurability of logic blocks is considered; the ALMmap tool is an example [66].

In recent years, research on system-oriented synthesis that focuses on System on Chip has become increasingly popular and is associated with the synthesis of Cyber-Physical Systems [14, 67, 68], where establishing a clear border between a physical layer of a system and implemented calculating techniques is difficult. Aspects related to the synthesis of energy-saving systems are also very important [69–73].

1.4 System on Chip

Technology development has enabled the manufacture of systems of once unimaginable complexity. This manufacture influences the expansion of the area of application of devices, such as hardware control systems [74–78], signal processing systems [79–82], hardware calculation methods [75, 83–85], hardware implementation of fuzzy systems [86–88], reconfigurable control systems concurrent [89–92], advanced data coding systems [80, 93, 94] and data analysis in transmission systems [95].

In most cases, the modern digital system is a microprocessor system that is surrounded by a peripheral digital system. This type of system is referred to as System on Chip. The programmable device is produced by merging various elements into one structure. The structure can be adapted to a user's needs, which is a characteristic feature of ASICs. One of the groups of systems on the chip, in which the main role is played by programmable structures, comprise the programmable systems on chip (programmable System on Chip (pSoC)). Formerly inaccessible systems implementation methods become real due to pSoC structures, which enable the building of dedicated complex devices, such as industrial PLC controllers [96–99].

Because these systems are currently developing dynamically, clearly classifying various solutions that appear on the market is difficult. The development of hardware description languages meant that complex elements of pSoC circuits (processors, arithmetic circuits) do not have to be created in the production process of an integrated circuit. Adaptation to needs can be achieved by programming the system. This possibility is created by virtual components, which are program resources of the system and resemble parameterized macros that were previously detected in synthesis tools. Thus, in pSoC systems, in addition to the programmable part, two types of additional resources can be distinguished: hardware resources produced in the integrated circuit production process and program resources implemented in the programming process, which are available in the form of various virtual components offered by many companies.

Programmable devices currently enable the development of hardware-software codesign systems that perform computational tasks in a multithreaded manner supported by hardware methods of thread switching and selection [100]. The problem of selection of virtual components, combined with issues of concurrent hardware and software synthesis and related to the design of these systems by methods of description (SystemC, SystemVerilog), is the basis for the development of circuits or programmable systems at the current time. A modern programmable system usually has the option of partial dynamic reconfiguration. This feature creates completely new possibilities for using the systems [101, 102] and introduces new problems into the synthesis process. In this type of programmable structure, for example, you can implement multicontext systems.

In recent years, the design process has been moved to areas very far from purely hardware issues. Designers describe systems at increasingly higher levels of abstraction, indicating that they often do not identify the hardware resources of the implemented system. According to the authors, this approach is not conducive to creating effective logic synthesis algorithms. This fact has become a motivation for writing this book. The development of programmable structures has constantly introduced new synthesis problems. Some of these problems have not been satisfactorily resolved. An example is the problem of decomposition. Despite the development of various decomposition methods, an effective method for dividing the designed system into configurable logic blocks contained in FPGAs does not exist. Showing circuits that use substantially fewer logic blocks than solutions obtained using commercial synthesis tools remains extremely easy [10, 11, 42, 103–108].

This book is an attempt to present a series of original concepts of technology mapping of digital circuits in FPGAs. The process of technology mapping is closely related to the process of decomposition of logic functions described by binary decision diagrams.

References

1. www.altera.com
2. www.xilinx.com
3. www.actel.com
4. Czerwiński R, Kania D (2009) Synthesis of finite state machines for CPLDs. Int J Appl Math Comput Sci (AMCS) 19(4):647–659
5. Czerwiński R, Kania D (2010) A synthesis of high speed finite state machines. Bull Polish Acad Sci Tech Sci 58(4):635–644
6. Kania D (2007) A new approach to logic synthesis of multi-output boolean functions on PAL-based CPLDs. In: Proceedings of the ACM great lakes symposium on VLSI, GLSVLSI'07 Stressa - Lago Maggiore, Italy, March, 11–13, 2007, pp 152–155
7. Kania D, Kubica M (2015) Technology mapping based on modified graph of outputs. In: International conference of computational methods in sciences and engineering, AIP Conference Proceedings, vol 1702
8. Opara A, Kania D (2010) Decomposition-based logic synthesis for PAL-based CPLDs. Int J Appl Math Comput Sci 20(2):367–384
9. Opara A, Kania D (2015) Logic synthesis strategy based on BDD decomposition and PAL-oriented optimization. In: 11th international conference of computational methods in sciences

and engineering, ICCMSE 2015, 20–23 March 2015, Athens, Greece, AIP Conf. Proc. 1702, 2015, 060002-1–4
10. Opara A, Kubica M (2016) Decomposition synthesis strategy directed to FPGA with special MTBDD representation. In: International conference of computational methods in sciences and engineering, American Institute of Physics, Athens, 17 Mar 2016, AIP conference proceedings, vol 1790
11. Cong J, Minkovich K (2007) Optimality study of logic synthesis for LUT-based FPGAs. IEEE Trans CAD 2(2):230–239
12. Kania D (2011) Efficient technology mapping method for PAL-based devices, design of digital systems and devices, book series. Lecture Notes in Electrical Engineering 79:145–163
13. Kania D (2015) Logic decomposition for PAL-based CPLDs. J Circuits Syst Comput 24(3):1–27
14. Opara A, Kubica M, Kania D (2019) Methods of improving time efficiency of decomposition dedicated at FPGA structures and using BDD in the process of cyber-physical synthesis. IEEE Access 7:20619–20631
15. Bolton M (1990) Digital systems design with programmable logic. Addison-Wesley Publishing Company, Boston
16. Chen SL, Hwang TT, Liu CL (2002) A technology mapping algorithm for CPLD architectures. In: IEEE international conference on field-programmable technology, Hong Kong, December 16–18, 2002, pp. 204–210
17. Czerwiński R, Kania D (2013) Finite state machine logic synthesis for CPLDs, Springer, Lecture Notes in Electrical Engineering, vol 231, XVI, 172 p
18. Kania D (2000) A technology mapping algorithm for PAL-based devices using multi-output function graphs. In: Proceedings of 26-th Euromicro Conference, IEEE Computer Society Press, Maastricht, 2000, pp 146–153
19. Kania D (2002) Logic synthesis of multi-output functions for PAL-based CPLDs. In: IEEE international conference on field-programmable technology, Hong Kong, December 16–18, pp 429–432
20. Kania D, Kulisz J, Milik A (2005) A novel method of two-stage decomposition dedicated for PAL-based CPLDs. In: Proceedings of Euromicro symposium on digital system design, IEEE Computer Society Press, Porto, September, pp 114–121
21. Espresso (1993) A source code. http://embedded.eecs.berkeley.edu/pubs/downloads/espresso/index.htm
22. Murgai R, Nishizaki Y, Shenay N, Brayton RK, Sangiovanni-Vincentelli A (1990) Logic synthesis for programmable gate array. In: Proceedings of 27th DAC, June 1990, pp 620–625
23. Murgai R, Shenoy N, Brayton RK, Sangiovanni-Vincentelli A (1991) Improved logic synthesis algorithms for table look up architectures. ICCAD-91, Santa Clara, CA, pp 564–567
24. Abouzeid P, Babba B, Crastes M, Saucier G (1993) Input-driven partitioning methods and application to synthesis on table-lookup-based FPGAs. IEEE Trans on CAD 12(7):913–925
25. Ashenhurst RL (1957) The decomposition of switching functions. In: Proceedings of an international symposium on the theory of switching, April 1957
26. Brown SD, Francis RJ, Rose J, Vranesic ZG (1993) Field programmable gate arrays. Kluwer Academic Publishers, Boston, pp 45–86
27. Brayton RK, Hachtel GD, McMullen C, Sangiovanni-Vincentelli AL (1984) Logic minimization algorithms for VLSI synthesis. Kluwer Academic Publishers, Boston
28. Brayton RK, Hachtel GD, Sangiovanni-Vincentelli AL (1990) Multilevel logic synthesis. Proc IEEE 78(2):264–300
29. Cong J, Ding Y (1994) FlowMap: an optimal technology mapping algorithm for delay optimization in lookup-table based FPGA design. IEEE Trans Comput-Aided Des 13(1):1–12
30. De Micheli G (1994) Synthesis and optimization of digital circuits. McGraw-Hill, Inc.
31. Chen D, Cong J (2004) DAOmap: a depth-optimal area optimization mapping algorithm for FPGA designs. Computer Aided Design, 2004. IEEE/ACM International Conference on ICCAD-2004, pp 752–759

32. Gowda T, Vrudhula S, Kulkarni N, and Berezowski K (2011) Identification of threshold functions and synthesis of threshold networks. IEEE Trans Comput-Aided Des Integr Circuits Syst 30(5)
33. Pistorius J, Hutton M, Mishchenko A, Brayton R (2007) Benchmarking method and designs targeting logic synthesis for FPGAs. In: Proceedings of Int'l workshop on logic and synthesis, vol 7
34. Curtis HA (1962) The design of switching circuits. D. van Nostrand Company Inc, Princeton
35. Curtis HA (1963) Generalized tree circuit—the basic building block of an extended decomposition theory. J ACM 10:562–581
36. Babba B, Crastes M, Saucier G (1992) Input driven synthesis on PLDs and PGAs. In: The European conference on design automation, Brussels (Belgium), March 1992
37. Huang J-D, Jou J-Y, Shen W-Z (2000) ALTO: an iterative area/performance tradeoff algorithm for LUT-based FPGA technology mapping. IEEE Trans Very Large Integration (VLSI) Syst 8(4):392–400
38. Legl Ch, Wurth B, Eckl K (1995) An implicit algorithm for support minimization during functional decomposition. ED&TC, Paris, pp 412–417
39. Pan KR, Pedram M (1996) FPGA synthesis for minimum area, delay and power, ED&TC, Paris, p 603
40. Wan W, Perkowski MA (1992) A new approach to the decomposition of incompletely specified multi-output functions based on graph coloring and local transformations and its applications to FPGA mapping. Proceedings of EDAC'92, pp 230–235
41. Kania D (2000) Decomposition-based synthesis and its application in PAL-oriented technology mapping. In: Proceedings of 26-th Euromicro conference, IEEE Computer Society Press, Maastricht, pp 138–145
42. Kubica M, Kania D (2016) SMTBDD: new form of BDD for logic synthesis. Int J Electron Telecommun 62(1):33–41
43. Perkowski M, Malvi R, Grygiel S, Burns M, Mishchenko A (1999) Graph coloring algorithms for fast evaluation of Curtis decompositions, 36-th ACM/IEEE DAC'99, New Orleans
44. Selveraj H, Łuba T, Nowicka M, Bignall B. Multiple-valued decomposition and its applications in data compression and technology mapping, ICCIMA'97
45. Jozwiak L, Chojnacki A (2003) Effective and efficient FPGA synthesis through general functional decomposition. J. Syst Archit 49(4–6):247–265. ISSN 1383-7621
46. Kania D, Kulisz J (2007) Logic synthesis for PAL-based CPLD-s based on two-stage decomposition. J Syst Softw 80:1129–1141
47. Kania D, Milik A (2010) Logic Synthesis based on decomposition for CPLDs. Microprocess Microsyst 34:25–38
48. Chang S, Marek-Sadowska M, Hwang T (1996) Technology mapping for TLU FPGA's based on decomposition of binary decision diagrams. IEEE Trans. Comput-Aided Des 15(10):1226–1235
49. Ebend R, Fey G, Drechsler R (2005) Advanced BDD optimization. Springer, Dordrecht
50. Kubica M, Opara A, Kania D (2017) Logic synthesis for FPGAs based on cutting of BDD. Microprocess Microsyst 52:173–187
51. Lai Y, Pedram M, Vrudhula S (1994) EVBDD-based algorithms for integer linear programming, spectral transformation, and function decomposition. IEEE Trans Comput Aided Des 13(8):959–975
52. Lai Y, Pan KR, Pedram M (1996) OBDD-based function decomposition: algorithms and implementation. IEEE Trans Comput-Aided Des 15(8):977–990
53. Machado L, Cortadella J, Support-reducing decomposition for FPGA mapping. IEEE Trans Comput-Aided Des Integr Circuits Syst 39(1):213–224
54. Opara A, Kubica M (2017) Optimization of synthesis process directed at FPGA circuits with the usage of non-disjoint decomposition. In: Proceedings of the international conference of computational methods in sciences and engineering 2017, American Institute of Physics, Thessaloniki, 21 Apr 2017, Seria: AIP Conference Proceedings, vol 1906, Art. no. 120004

55. Opara A, Kubica M, Kania D (2018) Strategy of logic synthesis using MTBDD dedicated to FPGA. Integr VLSI J 62:142–158
56. Opara A, Kubica M (2018) The choice of decomposition taking non-disjoint decomposition into account. In: Proceedings of the international conference of computational methods in sciences and engineering 2018, American Institute of Physics, Thessaloniki, 14 Mar 2018, AIP Conference Proceedings, vol 2040, Art. no. 080010
57. Sasao T (1993) FPGA design by generalized functional decomposition in logic synthesis and optimization. Kluwer Academic Publishers, Boston
58. Scholl A (2001) Functional decomposition with application to FPGA synthesis. Kluwer Academic Publishers, Boston
59. Zhang H, Chen Z, Wang P (2019) Area and delay optimization of binary decision diagrams mapped circuit. J Electr Inf Technol 41(3):725–731
60. Yang C, Ciesielski M (2002) BDS: a BDD-based logic optimization system. IEEE Trans Comput-Aided Des Integ Circuits Syst 21(7):866–876
61. Vemuri N, Kalla P, Tessier R (2002) A BDD—based logic synthesis for LUT—based FPGAs. ACM Trans Des Autom Electron Devices 7(4):501–525
62. Cheng L, Chen D, Wong MDF (2007) DDBDD: delay-driven BDD synthesis for FPGAs. In: Design automation conference, 2007, DAC '07. 44th ACM/IEEE, 2007, pp 910–915
63. Berkeley Logic Synthesis Group (2005) ABC: a system for sequential synthesis and verification, Dec. 2005. Available: http://www.eecs.berkeley.edu/~alanmi/abc
64. Fiser P, Schmidt J (2009) The case for a balanced decomposition process. In: Proceedings of 12th Euromicro conference on digital systems design (DSD), Patras (Greece), pp 601–604
65. Fiser P, Schmidt J (2012) On using permutation of variables to improve the iterative power of resynthesis, *in Proc. of 10th Int. Workshop on Boolean Problems (IWSBP)*, Freiberg (Germany), pp 107–114
66. Liang YY, Kuo TY, Wang SH, Mak WK (2010) ALMmap: technology mapping for FPGAs with adaptive logic modules, computer-aided design (ICCAD). In: 2010 IEEE/ACM international conference on, pp 143–148
67. Kubica M, Kania D, Kulisz J (2019) A technology mapping of FSMs based on a graph of excitations and outputs IEEE Access, vol 7, pp 16123–16131
68. Kubica M, Kania D (2019) Graph of outputs in the process of synthesis directed at CPLDs, Mathematics 7(12):1–17, Art. no. 1171
69. Kajstura K, Kania D (2011) A decomposition state assignment method of finite state machines oriented towards minimization of Power. Przegląd Elektrotechniczny 87(6):146–150
70. Kajstura K, Kania D (2016) Binary tree-based low power state assignment algorithm. In: 12-th international conference of computational methods in science and engineering, ICCMSE 2016, 17–20 March 2016, Athens, Greece, AIP Conference Proceedings 1790, 2016, pp 0300007_1–0300007_4
71. Kajstura K, Kania D (2018) Low power synthesis of finite state machines state assignment decomposition algorithm. J Circuits Syst Comput 27(3):1850041-1–1850041-14
72. Kubica M, Kajstura K, Kania D (2018) Logic synthesis of low power FSM dedicated into LUT-based FPGA. In: Proceedings of the international conference of computational methods in sciences and engineering 2018. American Institute of Physics, Thessaloniki, 14 Mar 2018, AIP Conference Proceedings, vol 2040
73. Mengibar L, Entrena L, Lorenz MG, Millan ES (2005) Patitioned state encoding for low Power in FPGAs. Electron Lett 41:948
74. Kobylecki M, Kania D (2017) FPGA implementation of bit controller in double-tick architecture. In: 13-th international conference of computational methods in science and engineering, ICCMSE 2017, 21–25 April 2017, Thessaloniki, Greece, AIP Conference Proceedings 1906, pp 120008_1–120008_4
75. Milik A, Hrynkiewicz E (2018) Hardware mapping strategies of PLC programs in FPGAs. In: 15th IFAC conference on programmable devices and embedded systems. PDeS 2018, Ostrava, Czech Republic, 23–25 May 2018. Amsterdam, Elsevier, 2018, pp 131–137

76. Mocha J, Kania D (2012) Hardware implementation of a control program in FPGA structures. Przegląd Elektrotechniczny R.88(12a):95–100
77. Wyrwoł B (2011) Using graph greedy coloring algorithms in the hardware implementation of the HFIS fuzzy inference system. Przegląd Elektrotechniczny, Warszawa, nr 10, R. 87:64–67
78. Ziebinski A, Bregulla M, Fojcik M, Klak S (2017) Monitoring and controlling speed for an autonomous mobile platform based on the hall sensor. In: Nguyen NT, Papadopoulos GA, Jędrzejowicz P, Trawiński B, Vossen G (eds) Computational collective intelligence: 9th international conference, ICCCI 2017, Nicosia, Cyprus, 27–29 Sept 2017, Proceedings, Part II. 2017, pp 249–259
79. Czerwinski R, Rudnicki T (2014) Examination of electromagnetic noises and practical operations of a PMSM motor driven by a DSP and controlled by means of field oriented control. Elektronika Ir Elektrotechnika 20(5):46–50
80. Pamula D, Ziebinski A (2009) Hardware implementation of the MD5 algorithm. In: Proceedings of programmable devices and embedded systems conference, Roznov pod Radhostem, pp 45–50
81. Pułka A, Milik A (2012) Hardware implementation of fuzzy default logic. In: Hippe ZS, Kulikowski JL (eds) Human-computer systems interaction backgrounds and applications II, Series: Advances in Intelligent and Soft Computing, vol 99. Springer, Berlin, pp 325–343
82. Rudnicki T, Czerwinski R, Sikora A, Polok D (2016) Impact of PWM control frequency onto efficiency of a 1 kW Permanent Magnet Synchronous Motor (PMSM). Elektronika Ir Elektrotechnika 22(6):10–16
83. Pułka A, Milik A (2011) An efficient hardware implementation of smith-waterman algorithm based on the incremental approach. Int J Electron Telecommun 57(4):489–496
84. Pułka A, Milik A (2012) Measurement aspects of the genome patterns investigations—hardware implementation. Metrol Meas Syst 19(1):49–62
85. Ziebinski A, Swierc S (2016) Soft core processor generated based on the machine code of the application. J Circuits Syst Comput 25(04):1650029
86. Wyrwoł B, Hrynkiewicz E (2013) Decomposition of the fuzzy inference system for implementation in the FPGA structure. In: Int J Appl Math Comput Sci 23(2):473–483
87. Wyrwoł B, Hrynkiewicz E (2016) Implementation of a microcontroller-based simplified FITA-FIS model. Microprocess Microsyst 44:22–27
88. Wyrwoł B (2019) Implementation of the FATI hierarchical fuzzy inference system using the immutability decomposition method. Fuzzy Sets and Systems. Elsevier, Amsterdam, pp 1–19
89. Grobelna I, Wisniewski R, Grobelny M, Wisniewska M (2017) Design and verification of real-life processes with application of petri nets. IEEE Trans Syst Man Cybern Syst 47(11):2856–2869
90. Wisniewski R, Bazydlo G, Gomes L, Costa A (2017) Dynamic partial reconfiguration of concurrent control systems implemented in FPGA devices. IEEE Trans Industr Inf 13(4):1734–1741
91. Wiśniewski R (2017) Prototyping of concurrent control systems implemented in FPGA devices. Springer International Publishing, Berlin
92. Wiśniewski R (2018) Dynamic partial reconfiguration of concurrent control systems specified by petri nets and implemented in Xilinx FPGA devices. IEEE Access 6:32376–32391
93. Sułek W (2016) Non-binary LDPC decoders design for maximizing throughput of an FPGA implementation. Circuits Syst Signal Process 35:4060–4080
94. Sułek W (2019) Protograph based low-density parity-check codes design with mixed integer linear programming. IEEE Access 7:1424–1438
95. Cupek R, Piękoś P, Poczobut M, Ziebinski A (2010) FPGA based "intelligent tap" device for real-time ethernet network monitoring, computer networks. Proc Ser: Commun Comput Inf Sci 79:58–66
96. Chmiel M, Czerwinski R, Smolarek P (2015) IEC 61131-3-based PLC Implemented by means of FPGA. In: 13th IFAC conference on programmable devices and embedded systems, PDeS'15, 13–15 May 2015, pp 383–388

97. Chmiel M, Kulisz J, Czerwinski R, Krzyzyk A, Rosol M, Smolarek P (2016) An IEC 61131-3-based PLC implemented by means of an FPGA. Microprocess Microsyst 44:28–37
98. Mazur P, Chmiel M, Czerwinski R (2016) Central processing unit of IEC 61131-3-based PLC. In: 14th IFAC conference on programmable devices and embedded systems, PDeS'16, 5–7 October 2016, pp 111–116
99. Nawrath R, Czerwiński R (2018) FPGA-based implementation of APB/SPI bridge. In: 14th international conference of computational methods in sciences and engineering, ICCMSE'18, Thessaloniki, Greece, 14–18 March 2018
100. Milik A (2018) Multiple-core PLC CPU implementation and programming. J Circuits Syst Comput 27(10)
101. Milik A (2016) On hardware synthesis and implementation of PLC programs in FPGAs. Microprocess Microsyst 44:2–16
102. Milik A, Pułka A (2011) Automatic implementation of arithmetic operation in reconfigurable logic controllers. In: Proceedings of ECCTD 2011 conference, Linköping, Sweden, Aug 27–30, pp 721–724
103. Kubica M, Kania D (2015) SMTBDD: new concept of graph for function decomposition. IFAC conference on programmable devices and embedded systems: PDeS
104. Kubica M, Kania D, Opara A (2016) Decomposition time effectiveness for various synthesis strategies dedicated to FPGA structures. In: 12th international conference of computational methods in science and engineering, ICCMSE 2016, 17–20 March 2016, Athens, Greece, AIP Conference Proceedings
105. Kubica M, Kania D (2017) Decomposition of multi-output functions oriented to configurability of logic blocks. Bull Polish Acad Sci Tech Sci 65(3):317–331
106. Kubica M, Kania D (2017) Area-oriented technology mapping for LUT-based logic blocks. Int J Appl Math Comput Sci 27(1):207–222
107. Kubica M, Milik A, Kania D (2018) Technology mapping of multi-output function into LUT-based FPGA. In: Slanina Z (ed) IFAC conference on programmable devices and embedded systems: PDeS 2018 Ostrava, Czech Republic, 23–25 May 2018. Elsevier, Amsterdam, 2018, pp 107–112
108. Kubica M, Kania D (2019) Technology mapping oriented to adaptive logic modules. Bull Polish Acad Sci—Tech Sci 67(5):947–956

Chapter 2
Methods for Representing Boolean Functions—Basic Definitions

In the process of logic synthesis, the method of representing Boolean functions has great importance. The development of synthesis methods is inseparably connected with the search for new forms of description [1–4]. A "good" description of logic expressions should be effective in terms of memory usage while ensuring easy optimization of Boolean expressions that are mapped in modern digital circuits.

This chapter focuses on several basic forms of function representation. These characters are directly related to the following considerations, which cover various issues of synthesis and technology mapping. Basic definitions are also presented, and they enable a precise description of various synthesis issues.

2.1 Hypercube, Cube, Implicant, Minterm

Let B^n be an n-dimensional binary space, which is often referred to as an n-dimensional cube or hypercube. A hypercube is a representation of an n-dimensional space, where each point is associated with n-binary coordinates [5].

Each point of the 4-dimensional space is associated with a unique vector that can be described by the product of 4 literals. A literal is a variable or its negation. For example, a point of space with coordinates 0101 corresponds to the product of four literals $\bar{d}c\bar{b}a$. The product of literals is sometimes referred to as a subcube or cube. If all elements of space appear in the product of literals, we refer to a 0 dimensional cube. Any subspace that is selected can be described by the product of literals. For example, in the 4-dimensional space represented by the hypercube in Fig. 2.1, one of the planes describes the product $\bar{d}a$. This product corresponds to the vector 0--1. In this case, we refer to a 2nd-order cube due to two elements of indeterminacy.

In the general case, the Boolean function f maps the set B^n into the set {0,1,-}, where B = {0,1}, i.e., f:f: $B^n \rightarrow$ {0,1,-}. Thus, each function can be uniquely defined by specifying two of three sets of B^n space points, for which the values 0, 1 and - are assumed. These sets are referred to as the OFF-set, ON-set and Don't Care (DC)-set.

M. Kubica et al., *Technology Mapping for LUT-Based FPGA*, Lecture Notes in Electrical Engineering 713, https://doi.org/10.1007/978-3-030-60488-2_2

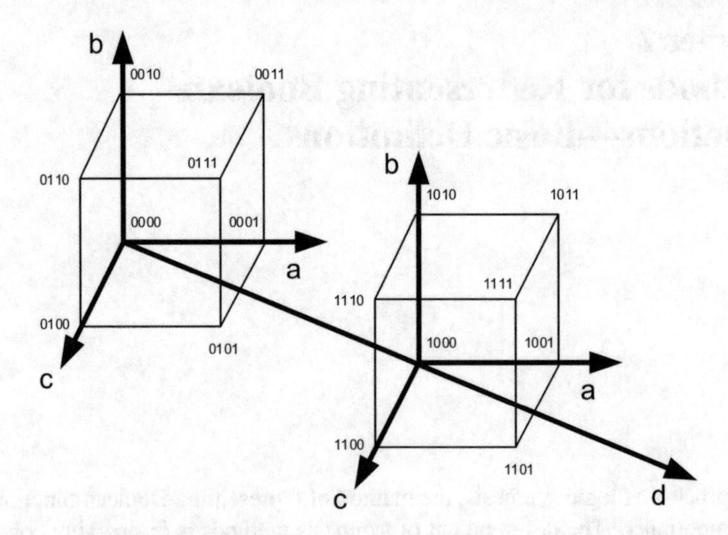

Fig. 2.1 Four-dimensional hypercube

Consider the logic function f: $B^4 \rightarrow B$, which is represented by a 4-dimensional hypercube. The values of the functions 0, 1, and - for individual coordinates are marked with the symbols o, •, and x, respectively (Fig. 2.2).

An unambiguous description of the function is possible by specifying two of the three subsets of the B^4 space, i.e., ON-set and OFF-set, ON-set and DC-set, or OFF-set and DC-set, where

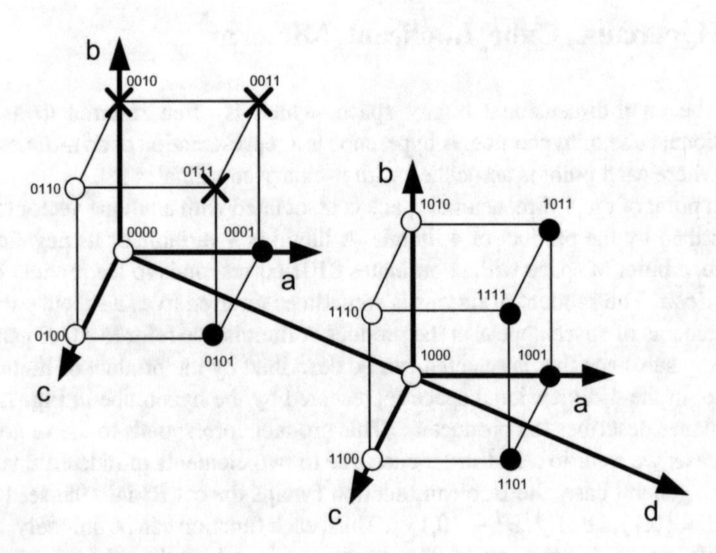

Fig. 2.2 Description of the example function f: $B^4 \rightarrow B^1$

$$\text{OFF} = \{0000, 0100, 0110, 1000, 1010, 1100, 1110\}_{dcba}$$
$$\text{ON} = \{0001, 0101, 1001, 1011, 1101, 1111\}_{dcba}$$
$$\text{DC} = \{0010, 0011, 0111\}_{dcba}.$$

Definition 2.1 An implicant of the Boolean function $f: B^n \to B$ is a pair of vectors with the dimensions n and 1, which are referred to as the input part and the output part. The input part contains elements of the set {0,1,-}, while the output part contains the element 1. The input part corresponds to the product of literals, for which the function takes the value 1 or DC.

Definition 2.2 A prime implicant of the Boolean function $f: B^n \to B$ is the implicant described by the product of literals, among which no literal can be removed and the reduced product remains the implicant of the function.

Definition 2.3 A minterm of the Boolean function $f: B^n \to B$ is the implicant (product of n literals), which in the input part contains only elements of the set {0,1}.

An example of the implicants of the function shown in Fig. 2.2 can be the pair of vectors 1--1 1 that correspond to the expression *ad*. The space points correspond to the implicant and lie on one plane, for which the variable *a* and *d* assumes the value 1. The implicant 1--1 1 covers the four minterms 1001 1, 1011 1, 1101 1 and 1111 1.

A very important issue is the simultaneous description of several functions. In the case of a multioutput function, the concepts of multioutput implicants and multioutput minterms have a significant role.

Definition 2.4 The multioutput implicant of the logic function $f: B^n \to B^m$ is a pair of vectors with the dimensions n (input part) and m (output part). The input part contains elements of the set {0,1,-}, and in the case of the implicant of a single function, represents the product of literals. The output part assumes the values of 0 or 1, where the value 1 is understood and ensures that the corresponding function for the vector that corresponds to the input part assumes the value 1 or indefinite value, while the value 0 does not contain any information about the value of the corresponding function.

Definition 2.5 The multioutput minterm of the Boolean function $f: B^n \to B^m$ is a multioutput implicant, which in the input part contains n-elements that belong to the set {0,1} (full product of n-literals) and in the output part contains only one 1.

We present in a 3-dimensional Boolean space a set of two functions *f0* and *f1*, which are described using three two-output terms: 11- 11, --1 10, and –01 01, where the order of the elements corresponds to the variables *cba* and the values of the function *f0f1*.

A certain problem that causes ambiguity is a misunderstanding of the symbol 0 in the output part of a multioutput implicant. This problem occurs in the case of the symbol 0 located in the output part of the multioutput implicant --1 10. This implicant entails four minterms of the function *f1*, i.e., 001 10; 011 10, 101 10, 111 10. However, the implicant does not contain any information about the function *f1*. For example,

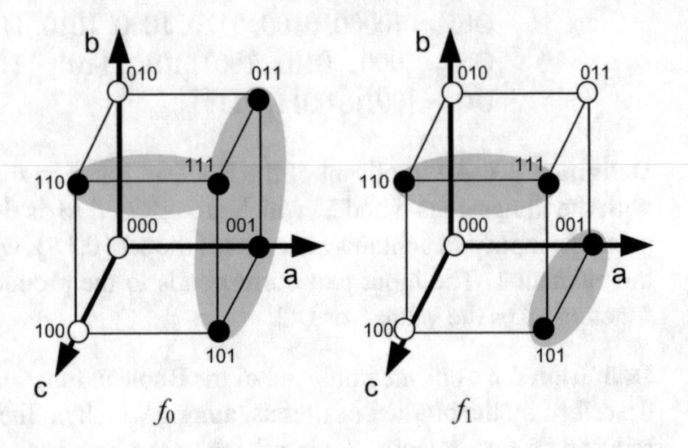

Fig. 2.3 Implicants of functions *f0* and *f1*

the symbol 0 in the output part that corresponds to the function *f1* of the multioutput minterm 001 10 does not mean that the function *f1* for the *cba* vector equal to 001 assumes the value 0. The value of this function defines the multioutput implicant -01 01, which does not contain any information about the function *f1* (Fig. 2.3).

Understanding the concept of the multioutput implicant and multioutput minterm enables a proper interpretation of a very popular description of functions in the Berkeley format used by the program to minimize Espresso [6–8].

2.2 Two-Level Description of a Boolean Function

One way to describe a Boolean function is the Berkeley format, which is accepted by the program to minimize Espresso [7, 8]. The function description file consists of three groups of lines:

- lines that contain comments starting with the character "#",
- lines that contain keywords that start with the character ".",
- lines that describe Boolean functions that start with one of the characters "0", "1" or "-".

Keywords determine the way in which each character is interpreted to form a description of the [9] function. The order in which the keywords appear is strictly defined. Their minimum set should contain the words ".i" and ".o", where: i[d] defines the number of inputs of the function, .o[d] is the number of outputs of the function, where d represents the corresponding decimal number. Labels can be assigned to individual inputs and outputs using the following structures: .ilb [s1]… [sn] - assign symbolic names to all variables of the function; and .ob [s1]… [sn] - assign symbolic names to all functions, where [si] is the name of the corresponding function input or output. The description of logic functions is supplemented by two additional keywords: .p [d] - specifies the number of lines that form the description of the function (d—decimal number) and (.e (.end) - end of description marker). Details of the Espresso format for the function description are provided in [9].

As previously mentioned, the unambiguous description of the function $B^n \to \{0, 1,-\}^m$ enables two of three disjoint subsets of B^n spaces to be specified, for which the function assumes the values of 0, 1 and—are referred to as the On-set, OFF-set and DC-set, respectively. If the ".type" keyword is not included in the function description file, the function description should be interpreted to include the description ON-set and DC-set, and OFF-set, which can be designated as the complement of the sum of specific sets. This interpretation affects the way characters are interpreted in the lines that describe the implications of the functions, where the symbol 1 means that the cube belongs to the ON-set, the symbol - indicates that the cube belongs to the DC-set, and the symbol 0 should be interpreted that the cube does not belong to the description of the function.

Thus, each line in the descriptive part of the function $B^n \to \{0,1,-\}^m$ consists of the n element input part, and this part contains the elements {0,1,-} and the m element output part, which consists of the elements {0,1}. In the case of a description of the function sets, some n-element input parts can be used to describe at least one function. The information in the output part, of which the cube does not belong to the description of the selected function, enables significant compression of the description. An example of a 3-inputs description and 2-outputs function is presented in Fig. 2.4.

The extension of the two-level description of a Boolean function is the description of sequential automata in the form of the KISS format. In this case, in the place of multioutput implicants, the following lines contain a symbolic description of FSM transitions, which are sometimes referred to as symbolic implicants. An example of a symbolic implicant can be the line 011 s1 s2 01, which means that an FSM in the state s1 at the input vector 011 goes to the state s2 and simultaneously sets the FSM

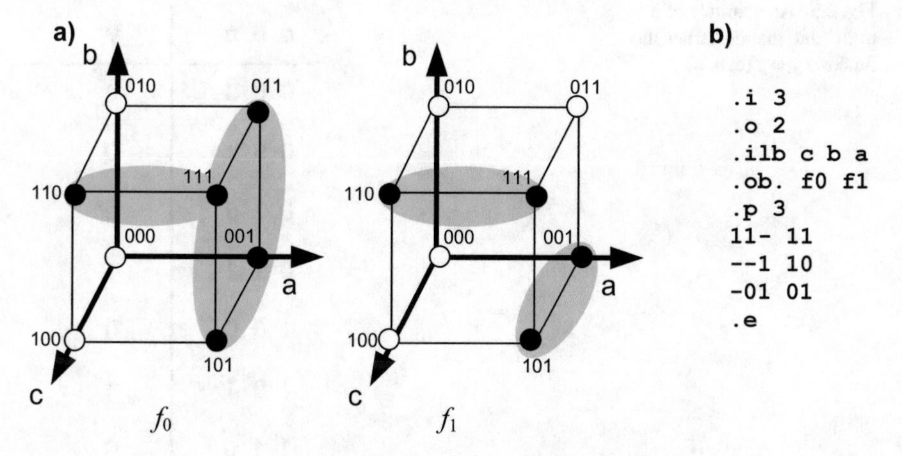

Fig. 2.4 An example of the description of the function $B^3 \to B^2$ in the form of hypercube with marked terms (**a**) and the corresponding file in the form * .pla (**b**)

outputs 01 on the FSM outputs. Details of the FSM description in KISS format are provided in the [10].

2.3 Truth Table, Karnaugh Map, Binary Decision Tree

One of the original forms of describing Boolean functions is the truth table. The description of a 3-input function using a truth table requires the creation of an array of 2^3 rows associated with subsequent combinations of variables and their corresponding function values, as shown in Fig. 2.5.

Many forms of the description of logic functions, from the point of view of logic synthesis, do not have any practical significance. Imagine the size of the truth table that would be used to describe a 16-output combinational system implemented in a small 100-I/O programmable device. If we wanted to map this table in the computer's memory, we would need to use $2^{84} * 16$ memory cells. These types of values exceed the available memory capacities many times.

The development of synthesis methods is inseparably connected with the search for additional effective ways to describe logic functions [4]. In the initial period of development of digital technology, the basic problem of synthesis is the minimization of logic expressions [11]. The classic goal of logic minimization is to reduce expressions that comprise the sum of products. This minimization goal is closely related to the problem of implementing digital circuits using a gate network. The resulting network of gates should contain the minimum number of gates that have the smallest possible number of inputs. This network directly translates to a minimization of integrated circuits and the number of soldering points, reducing the cost of a printed

Fig. 2.5 An example of a truth table that describes the function $y = f(c, b, a)$

c b a	y
0 0 0	0
0 0 1	0
0 1 0	0
0 1 1	1
1 0 0	0
1 0 1	1
1 1 0	0
1 1 1	1

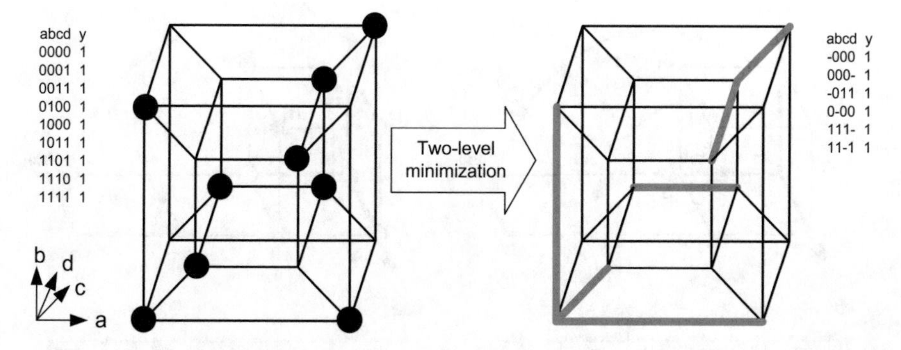

Fig. 2.6 Two-level minimization

circuit board and increasing the reliability of the entire device. Currently, circuits are not implemented in the form of a network of gates associated with separate integrated circuits. However, minimization remains extremely important, especially in the case of mapping logic functions in gate circuits, e.g., CPLD. The initial development of minimization methods is directly related to the representation of the Boolean function in the form of the Karnaugh map [12].

Let f be a logic function that maps the set B^n to set B, i.e., f: $B^n \rightarrow B$, where set B includes two values of B = {0,1}. Each function can be unequivocally described by a set of mintems that comprise a set of vectors with the dimensions n and 1, which are referred to the input part and the output part, respectively. The input part consists of 0, 1 elements and represents the full product of literals, for which the function assumes the value 1, which is the output part of a single-output minterm.

The function f: $B^3 \rightarrow B$, which is described by a set of single-output minterms associated with individual points of the four-dimensional Boolean space (Fig. 2.6), can be minimized and presented in a simpler form. Two-level minimization consists of covering the points of four-dimensional space (marked with black circles), which comprise a set of a minimal number of implicants. The results of the minimization for the considered function was presented on the right side Fig. 2.6.

The goal of minimization can be achieved by various methods. The original idea of presenting the n-dimensional Boolean space on a plane was proposed by Karnaugh. The idea is to replace the vertices of the hypercube with boxes, where you can enter the appropriate function values. The idea of presenting an n-dimensional space on a plane enabled the development of an effective minimization algorithm using graphical relationships that are illustrated using the Karnaugh map. The development of a three-dimensional space on a plane and a simple example of the idea of minimization is shown in Fig. 2.7.

Analysis of the value of the example function associated with the respective vertices of the hypercube shows that the 1 values (black points) are located on one of two lines. The first line traverses points 011 and 111, and the second line traverses 101, 111. The result of minimization is described by the function f (c, b, a) = ab + ac. An identical conclusion can be drawn by analyzing the system of 0 and

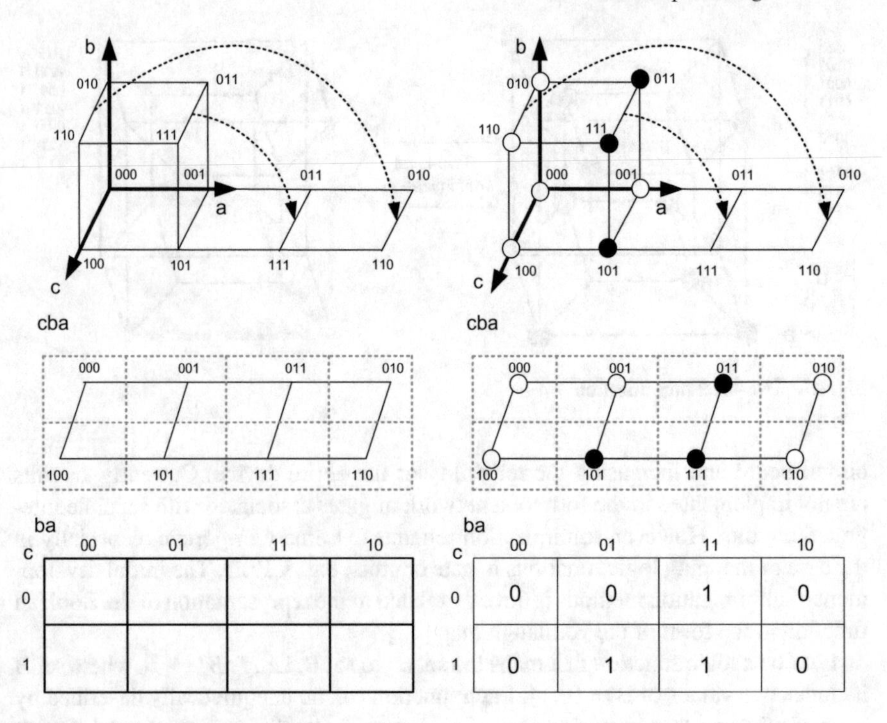

Fig. 2.7 Minimization using the Karnaugh map

1 in the 3-dimensional Karnaugh map. The use of geometrical properties of the function description by means of the Karnaugh map enables minimization of the function, which involves searching for a minimum number of the largest groups of ones that are symmetrical with respect to the appropriate number of axes [13]. The method of minimizing functions using the Karnaugh map can be practically employed only for functions for which the number of variables is small, practically no more than six. Theoretically, this method can also be applied to describe and minimize the multioutput function, which is an immense problem. Therefore, the practical significance of Karnaugh maps is currently negligible.

Binary decision diagrams, which will be presented in the next chapter, are forms of description that have been very promising since the end of the last century. The original form of describing the functions from which binary decision diagrams originate are binary decision trees. The nodes of this tree are associated with the variables of the function. Each node has two edges that correspond to the values of variables 0 and 1. The number of levels of the binary decision tree is equal to the number of variables in the function. Each level has one variable. The tree leaves specify the function values. Each leaf is associated with one unique combination of function variable values that describes the path from the root of the binary decision tree to the selected leaf. Each path from the root of the binary decision tree to the selected leaf corresponds to the minterm of the function. A translation of three equivalent

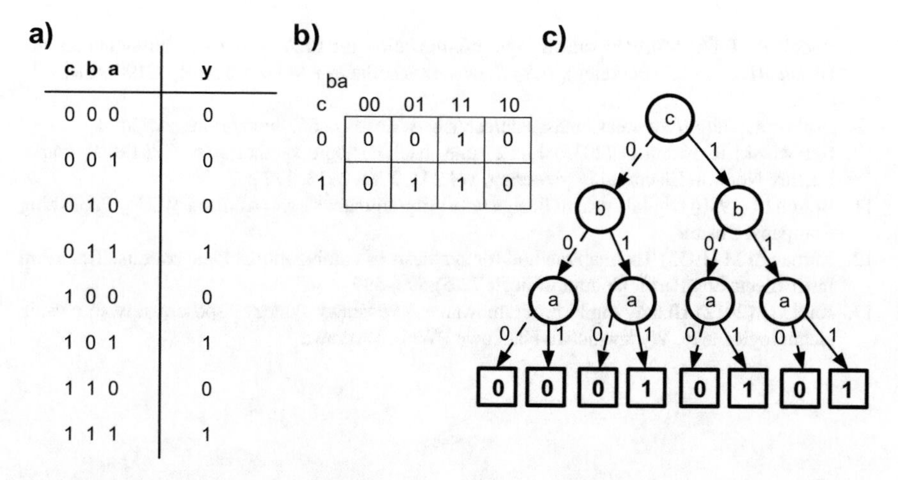

a)

c b a	y
0 0 0	0
0 0 1	0
0 1 0	0
0 1 1	1
1 0 0	0
1 0 1	1
1 1 0	0
1 1 1	1

b)

c \ ba	00	01	11	10
0	0	0	1	0
1	0	1	1	0

Fig. 2.8 Methods for representing the function $f: B \rightarrow B^1$, in the form of a truth table (**a**), Karnaugh map (**b**) and binary decision tree (**c**)

representations of the function $f: B \rightarrow B$ in the form of a truth table, Karnaugh map and binary decision tree is shown in Fig. 2.8.

The total number of binary decision tree nodes (sum of the number of nodes associated with function variables and sum of leaves) is $2^{n+1} - 1$ (n-number of function variables), which means that this representation is not effective in terms of memory usage. However, the possibility of a significant reduction in the number of nodes enables the possibility of applying the rules of elimination and joining of nodes.

The function description methods in this chapter have numerous disadvantages, explaining why graphical methods of function description are often employed in practical applications. An effective method of function representation is the use of binary decision diagrams, as shown in Chap. 3.

References

1. Akers SB (1978) Binary decision diagrams. IEEE Trans Comput C-27(6):509–516
2. Bryant R (1986) Graph based algorithms for boolean function manipulation. IEEE Trans Comput C-35(8):677–691
3. Kubica M, Kania D, Kulisz J (2019) A technology mapping of FSMs based on a graph of excitations and outputs. IEEE Access 7:16123–16131
4. Micheli G (1994) Synthesis and optimization of digital circuits. McGraw-Hill, New York
5. Ashar P, Devades S, Newton A (1992) Sequential logic synthesis. Kluwer Academic Publishers, Berlin
6. Brayton R, Hachtel G, McMullen C, Sangiovanni-Vincentelli A (1984) Logic minimization algorithms for VLSI synthesis. Kluwer Academic Publishers, Berlin
7. Espresso (1993) A source code. http://embedded.eecs.berkeley.edu/pubs/downloads/espresso/index.htm

8. Rudell R (1986) Multiple-valued logic minimization for PLA synthesis, Memorandum No. UCB/ERL M86-65 (Berkeley). (http://www.eecs.berkeley.edu/Pubs/TechRpts/1986/ERL-86-65.pdf
9. EspFormat. http://www.ecs.umass.edu/ece/labs/vlsicad/ece667/links/espresso.5.html
10. Czerwiński R, Kania D (2013) Finite state machine logic synthesis for CPLDs, Springer, Lecture Notes in Electrical Engineering, vol 231, XVI, 2013, 172 p
11. Bolton M (1990) Digital systems design with programmable logic. Addison-Wesley Publishing Company, Boston
12. Karnaugh M (1953) The map method for synthesis of combinational logic circuits. Trans Am Instit Electr Eng Part I: Commun Electr 72(5):593–599
13. Kania D (2012) Układy logiki programowalne – Podstawy syntezy i sposoby odwzorowania technologicznego, Wydawnictwo Naukowe PWN, Warszawa

Chapter 3
Binary Decision Diagrams

An effective way to represent logic functions is the Binary Decision Diagram (BDD). The prototype of the BDD was the Binary Decision Programs (BDP) [1], in which different nodes in a given path can be assigned the same variable. The model proposed by Lee, which is also investigated by Akers, has not generated extensive interest for a long time. By adding restrictions on the order of variables and introducing reduction mechanisms, Bryant substantially increased the performance of algorithms using diagrams in 1986 [2]. The key is to introduce a definition of a BDD.

Definition 3.1 A Binary Decision Diagram (BDD) is a directed, acyclic, and connected graph (tree directed to the leaves) that is built of nodes and the directed transitions between them. Each node of this graph is associated with one function variable. Each node has two edges associated with the values of the variables 0 and 1. Each node of the graph (except the root) has at least one edge from nodes at a higher level. The binary decision diagram contains two terminal nodes (leaves) that are associated with the values of the function 0 or 1. The analysis of paths that occur in the BDD enable the values of individual variables, for which the function assumes a value of 1 or 0, to be determined.

In practice, only ordered BDD and reduced ordered BDD are used.

Definition 3.2 An Ordered Binary Decision Diagram (OBDD) is a BDD diagram, where the variables occur in the same order in each path from root to leaf and the given variable in the path occurs at most once. Some nodes are associated with the same variable at the given level of the diagram.

Definition 3.3 A Reduced Ordered Binary Decision Diagram (ROBDD) is an OBDD diagram that contains a minimum number of nodes for a given variable order. This diagram can be obtained by merging the appropriate nodes and reducing identical subdiagrams in the OBDD diagram.

A graphical representation of the diagram for an example function is shown in Fig. 3.1.

M. Kubica et al., *Technology Mapping for LUT-Based FPGA*, Lecture Notes in Electrical Engineering 713, https://doi.org/10.1007/978-3-030-60488-2_3

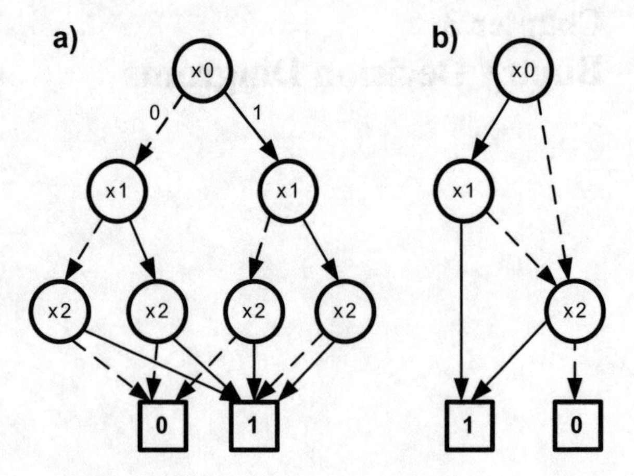

Fig. 3.1 Diagram that describes the function: $f = x_0 \cdot x_1 + x_2$, **a** OBDD, **b** ROBDD

In the diagram, the edges represent the logic values 0 and 1, which correspond to the Shannon expansion of the logic function:

$$f = \bar{x}_i \cdot f_{\overline{x_i}} + x_i \cdot f_{x_i} \tag{3.1}$$

where

- i node index,
- $f_{\bar{x}_i}$ function indicated by the node edge for the value 0 $f_{\overline{x_i}} = f_{x_i=0} = f(x_0, x_1, \ldots, x_{i-1}, 0, x_{i+1}, \ldots, x_n)$,
- f_{x_i} function indicated by the node edge for the value 1 $f_{x_i} = f_{x_i=1} = f(x_0, x_1, \ldots, x_{i-1}, 1, x_{i+1}, \ldots, x_n)$.

For the given node **v** associated with the variable x_{i}, which is the root of the diagram of the function f, the node indicated by the edge $x_{\text{i}} = 0$ from **v** is marked as low(**v**). Similarly, high(**v**) means the node indicated by the edge $x_{\text{i}} = 1$. The nodes low(**v**) and high(**v**) are descendants of the node **v** and the roots of the subdiagrams of the functions $f_{\overline{x_i}}$ and f_{x_i}, respectively.

The solid line draws the edges for the value 1 and the dotted line for the value 0. To obtain the ROBDD diagram, use the following reduction rules [2]:

1. Elimination rule—a node whose edges point to an identical node should be removed (Fig. 3.2a),
2. Merge rule—identical subdiagrams should be replaced by one subdiagram (Fig. 3.2b).

An important feature of the ROBDD diagrams is their canonicality, i.e., for each fixed order of variables, the ROBDD diagram of a logic function is clearly defined.

In practical implementations, an unreduced diagram is rarely created, and in the next step, it is reduced - this procedure would require significant memory and time complexity of algorithms. Libraries of subroutines that operate on diagrams already at the stage of creating each node ensure its uniqueness, and consequently, a lack

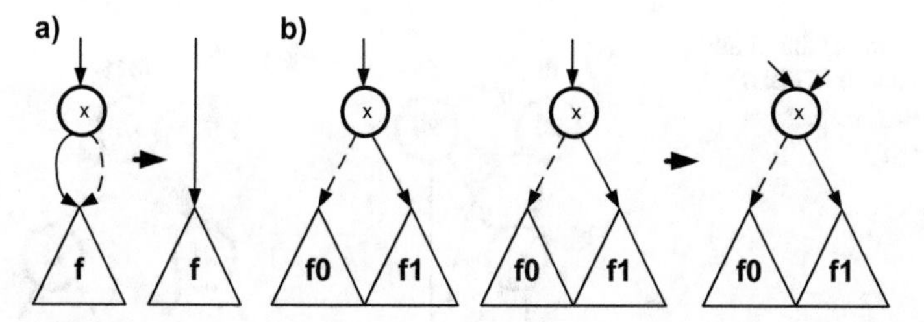

Fig. 3.2 Reduction rules: **a** elimination rule, and **b** merge rule

of repetitive subdiagrams. In addition, the first reduction condition is also checked when it occurs, and a new node is not created.

To reduce the computation time and memory needed to represent the diagram, attributes assigned to the edge [3] have been introduced in some diagrams. An example of this attribute is negation. This attribute means that you must negate the function that is represented by the subdiagram indicated by the edge with the attribute. A more extensive discussion of the use of this attribute is provided in this chapter.

Another known attribute is e.g., offset [3]. For example, given the functions $f_0(x_0, x_1, x_2) = x_0 \cdot x_1 + x_1 \cdot x_2$, $f_1(x_1, x_2, x_3) = x_1 \cdot x_2 + x_2 \cdot x_3$, they can be represented in memory by the same set of nodes, with the exception that the edge with the index shift attribute will be used to represent f_1. This attribute specifies how the quantity to add to the variable indexes of the function that is represented by the subdiagram indicated by the edge with the attribute.

Other examples of BDD variants are defined as follows:

- Free BDD (FBDD)—diagram without restrictions on the identical order of the variables in the paths, only with the requirement that the variable appears once in each path. These diagrams are used to reduce the number of memory cells needed to represent the function [4].
- Zero-suppressed BDD (ZBDD)—diagram with the elimination rule that consists of removing nodes for which the edge corresponds to 1 leads to leaf 0. If no node in the path associated with the variable x_i exists, then for $x_i = 1$ the function assumes the value 0. These diagrams are well suited for representing the combination [5].
- Ordered Functional Decision Diagram (OFDD)—a diagram in which the Reed-Muller (Davio) expansion is applied instead of Shanon's expansion. These diagrams have the ability to quickly perform disjointed sum operations (XOR) but are not very effective for the product and sum operations [6].
- Edge-Valued BDD (EVBDD)—diagram with assigned weights to the edges. These diagrams are suitable for representing arithmetic functions with integer values [7].

Variations of the diagrams enable a set of several functions to be effectively represented in terms of the number of memory cells. A multioutput function can be represented by one multiroot diagram (Fig. 3.3a). The functions are subdiagrams;

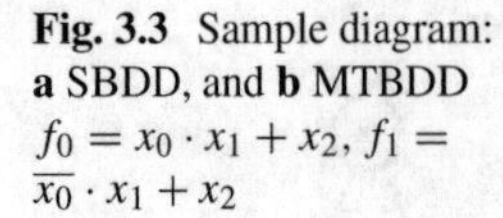

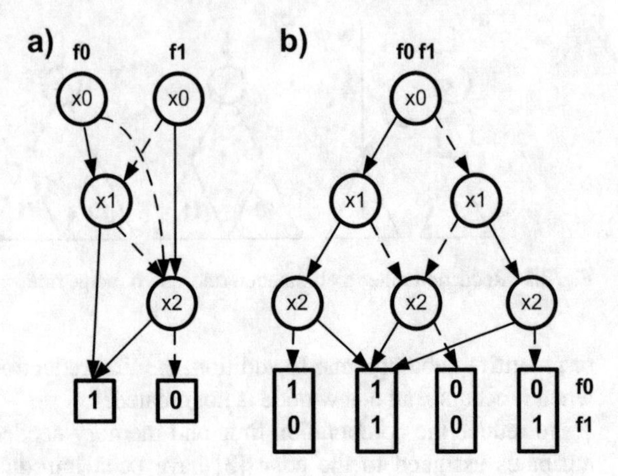

Fig. 3.3 Sample diagram: **a** SBDD, and **b** MTBDD $f_0 = x_0 \cdot x_1 + x_2$, $f_1 = \overline{x_0} \cdot x_1 + x_2$

thus, this structure is referred to as the Shared BDD (SBDD). The advantage of this solution is the concise representation and easier testing of the equivalence of the two functions.

Another way to represent a multioutput function is to use multileaf diagrams (Multi-Terminal BDD-MTBDD) [8]. In these diagrams, one root is common to all functions while each leaf contains a vector of m logic values, where m is the cardinality of the set of functions (Fig. 3.3b). The logic value vector is often presented in decimal form.

The presented BDD variants do not exhaust all proposals in the literature but only suggest the extent of the problem. In particular, ways to represent the DC function do not exist [3].

The size of the ROBDD that is represented by as the number of nodes depends on the order of variables in the diagram [9]. The strength of this relationship is indicated by the example of the function $f(x_1, \ldots, x_{2n}) = x_1{\cdot}x_2 + x_3{\cdot}x_4 + \cdots + x_{(2n-1)} \cdot x_{2n}$, for which the ROBDD size for order $x_1, x_2, \ldots, x_{2n}$ linearly increases with an increase in n, while the increase is exponential for ordering $x_1, x_3, \ldots, x_{(2n-1)}, x_2, x_4, \ldots, x_{2n}$. Finding the order of variables for which the ROBDD size is the smallest is an NP-hard problem [10]. The optimal solution algorithms were proposed in [11, 12]. Due to the time complexity, only heuristic algorithms are employed in practice, which enables the BDD diagram with the smallest possible number of nodes to be identified (e.g., using the 'divide and conquer' technique [13]). Algorithms that obtain an approximate solution can be divided into algorithms that achieve a reasonable ordering of variables before a diagram [14] is built and algorithms that change the order in a previously built diagram [3, 15, 16]. The latter includes, the window permutation algorithm [17], sifting algorithm [18] block sifting [19] and many modifications of algorithms that involve swapping the order of adjacent variables in swap order.

3.1 BDD Operations

To calculate the results of any binary logic operation (e.g., OR, and AND) for the two functions f and g presented in the form of BDD, the relationship 3.2 can be employed. The symbol * denotes any binary logic operation.

$$f * g = \bar{x}_i \cdot \left(f_{\overline{xi}} * g_{\overline{xi}}\right) + x_i \cdot (f_{xi} * g_{xi}) \tag{3.2}$$

The basic idea of performing BDD operations is to recursively apply the dependence to each node, starting from the root in the direction of the leaves [4]. At each recursion step, the node **v** with the assigned variable x_i is created, for which low(**v**) is the root of the function diagram $\left(f_{\overline{x_i}} * g_{\overline{x_i}}\right)$, while high(**v**) is the function $(f_{xi} * g_{xi})$. For the given procedure for functions with n variables to calculate $(f * g)$, 2^n subroutine calls are needed. The variables for each subroutine call are the subdiagrams of the functions *f* and *g*. The number of subdiagrams in the reduced diagram is equivalent to the number of nodes in the reduced diagram for the mentioned functions but their number is substantially smaller than 2^n in practice. Thus, the conclusion is that many of the 2^n calls can be avoided. Suppressing subroutine calls is possible due to the use of a cache memory that is referred to as the results table (Algorithm 3.1). The results table stores the calculation results for all encountered pairs of subroutine variables. Due to this solution, the computational complexity of the algorithm for constructing the function diagram (*f* * *g*) derived from any binary operation * is limited to O(sizeBDD(*f*) · sizeBDD(*g*)).

Algorithm 3.1: Binary operation on the BDD diagram

```
1.  calculate (f, g, *) {
2.  / * f, g - function diagrams, * - binary operator * /
3.  / * received diagram (f * g) * /
4.  if (f and g are leaves) { return (f * g); }
5.    else if ((f, g) ∈ ResultTable) { return (ResultTable (f, g)); }
6.    else {
7.    xi = first significant variable for f or g;
8.    v = new node;
9.    high (v) = calculate (fxi = 1, gxi = 1, *);
10.   low (v) = calculate (fxi = 0 gxi = 0, *);
11.   InsertToResultTable (f, g, v);
12.   return (v);
13.   }
14. }
```

To standardize the performance of all logic operations, Brace, Rudell and Bryant introduced the ITE operator (If-Then-Else) [20], which is defined according to expression (3.3).

$$\text{ITE}(x, y, z) = x \cdot y + \bar{x} \cdot z \tag{3.3}$$

Each of the binary operations can be presented using the ITE operator, e.g.,: $f \cdot g = \text{ITE}(f, g, 0)$, $f + g = \text{ITE}(f, 1, g)$, $f \oplus g = \text{ITE}(f, \bar{g}, g)$, and $\bar{f} = \text{ITE}(f, 0, 1)$. Similar to calculating the results of the binary operations, ITE can also use a recursive results calculation.

$$\text{ITE}(f, g, h) = \text{ITE}\big(x_i, \text{ITE}(f_{xi}, g_{xi}, h_{xi}), \text{ITE}\big(f_{\overline{xi}}, g_{\overline{xi}}, h_{\overline{xi}}\big)\big) \tag{3.4}$$

Algorithm 3.2: If-Then-Else operation on a BDD diagram

```
1.  ITE (f, g, h) {
2.  / * f, g, h - function diagrams * /
3.  / * ITE diagram (f, g, h) is obtained * /
4.  if (terminal case) { return (result of the terminal case); }
5.  else if ((f, g, h) ∈ ResultTable) { return (ResultTable (f, g, h)); }
6.  else {
7.    xi = first significant variable for f, g or h;
8.    t = ITE (f_xi = 1, g_xi = 1, h_xi = 1);
9.    e = ITE (f_xi = 0 g_xi = 0, h_xi = 0);
10.   r = FindOrAddToUniqueTableNodes (i, t, s);
11.   InsertToResultTable ((f, g, h), r);
12.   return (r);
13. }
14. }
```

Recursion is stopped when the ITE operator's arguments are constants, and in these cases, when the operator's result is one of the arguments, e.g., ITE $(1, f, g) = f$, ITE $(f, 1, 0) =$ f, ITE $(0, f, g) = g$, and ITE $(f, g, g) = g$. The algorithm for calculating ITE operations is presented in algorithm 3.2. Similar to algorithm 3.1, the Results Table is used to store the results of previous calculations. The new element is a Unique Table Nodes with pointers to all nodes. A new node is created only if no created node is associated with the same variable x_i, with edges that point to the same descendants. Because a given node can arise from ITE operations with different arguments, searching the Results Table is not sufficient for ensuring that a given node is unique; searching the table of unique nodes is necessary. Searching the array of unique nodes requires the merge rule. If you add the condition in the algorithm that the new node is not created when both outgoing edges point to the same node, then the diagram will be reduced when it is created.

When calculating the results of ITE operations for three—f, g, and h—diagrams, the arguments for subsequent calls to the ITE subroutine are the subdiagrams of f, g, and h. The number of subdiagrams in the diagrams is equivalent to the number of nodes in the diagrams. For each of the three arguments, the subroutine is called at most once. Assuming that the search in the Results Table is performed at a constant time, the computational complexity of the ITE algorithm, even in a pessimistic case, does not exceed O(sizeBDD(f) · sizeBDD(g) · sizeBDD(h)).

3.2 Basics of Software Implementation

In addition to the theoretical effectiveness of the algorithms, their practical implementation is very important. Currently, several different subprogram libraries are available to perform logic operations on functions represented by BDD [21]. These solutions do not need to be created from scratch. Common features of the most used libraries are presented in the next section.

3.2.1 Representation of Diagram Nodes in Computer Memory

Key progress in the effective implementation of BDD operations was achieved by Brace, Rudell and Bryant [20]. The programming techniques proposed by these researchers are currently referenced in the current libraries. Each node in the diagram is represented by an appropriate structure in memory (Fig. 3.4). The field marked index is the index of the variable x_i, which is the node that is being labeled. The low field and high field are pointers to descendants for edges $x_i = 0$ and $x_i = 1$ (solutions without pointers [22] are also described in the literature).

Due to the use of an array of unique nodes, the two equivalent functions f and g not only have the same reduced form of BDD but also are represented by a pointer to the same memory location in implementation. Testing the equivalence of two functions is then limited to a comparison of the values of two pointers.

Each node can be uniquely identified by three parameters: the variable index x_i and pointers to two descendants. Information regarding a node characterized by a tuple has been created, (index, low, and high) is stored in an array of unique nodes. To enable quick access to the data contained in the table, the data are implemented as a hash table [20, 23]. In classic methods, the search consists of comparing the

Fig. 3.4 Structure fields that represent the diagram node

Field	Type
Index	int
Low	ptr
High	ptr
Next	ptr
ref_count	short
Mark	short

searched key with elements of the searched set. Hashing attempts to obtain a direct reference to elements in an array using arithmetic operations that transform the key into the appropriate address of the array. The simplest hash function h can be the remainder of division:

$$h(index, low, high) = (index + low + high) \bmod Table\_Size \tag{3.5}$$

For example, for the $\mathbf{v}_2$ node (associated with x_1) with the $\mathbf{v}_5$ and $\mathbf{v}_7$ descendants (Fig. 3.5a), the hash function value is $(1 + 5 + 7) \bmod 8 = 5$. The calculated hash function value serves as an index in the table of unique identifiers. In a cell of table number 5, a pointer to node $\mathbf{v}_2$ exists (Fig. 3.5b). If no pointer to the node in the array cell with the calculated index exists, no node has yet been created. Hash functions usually do not have a different value, which causes a collision. For example, for node $\mathbf{v}_3$, we obtain the index in the Table $(2 + 7 + 6) \bmod 8 = 7$; for $\mathbf{v}_4$ and $\mathbf{v}_5$, we also obtain the same value 7. The next field in the structure of Fig. 3.4 is used to solve the collision problem and create a list of nodes for which the hash function value is identical. In Fig. 3.5b, in table cell number 7 a pointer to node $\mathbf{v}_3$ exists, and the next field of node $\mathbf{v}_3$ points to $\mathbf{v}_4$. In practice, slightly more complicated hash functions are employed but the idea of collision resolution remains the same.

Recently created libraries use one array of unique identifiers for each level of the diagram (for each variable index). Identifier management is then facilitated when changing the order of variables in the diagram. Access to all nodes from a given level is possible by browsing all lists, whose initial elements are only listed in one table.

The mark field in the structure that describes the node acts as a temporary marker in some algorithms, e.g., to indicate which nodes have been "visited", i.e., for which the algorithm does not need to be executed.

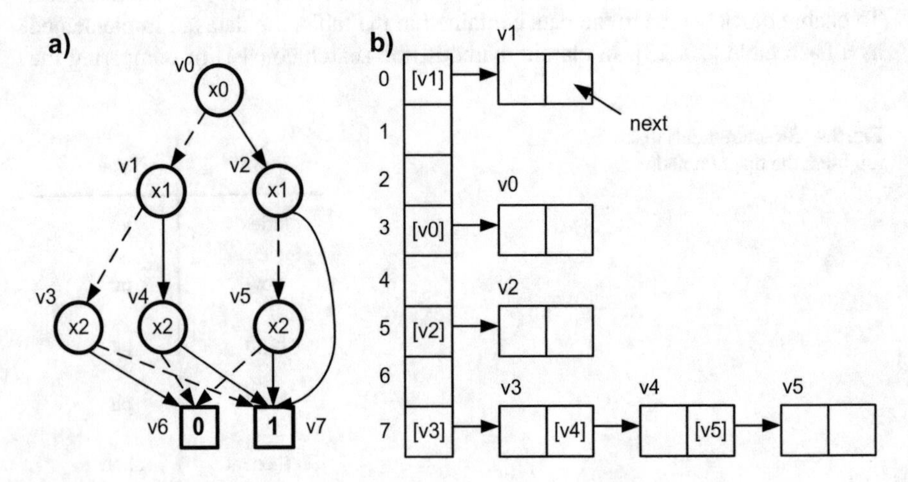

Fig. 3.5 Example of a diagram and its corresponding array of unique identifiers

The ref_count field indicates whether the node is currently being used. As soon as a node is created, the value of this field is 1. This value corresponds to the number of edges that point to a given node and is incremented (decremented) when creating (deleting) other nodes with edges that point to a given node. When the ref_count value reaches 0, the node is considered dead. However, this node is not immediately removed. Deleting a node each time would be too time-consuming. In addition, information about the existence of this node is included in the results table, and during subsequent calculations, this node can be utilized. For these reasons, instead of immediately freeing memory, waiting until a large number of dead nodes and then freeing unused memory areas comprises a better approach. This strategy is referred to as garbage collection [20, 24]. When collecting garbage, in addition to releasing memory allocated to the structure that represents a node, the entry in the table of unique identifiers must also be released and entries in the results table point to the removed node. Computational overhead related to freeing memory is only profitable when the free space for new nodes is not sufficient. Another reason for garbage collection may be to overfill the array of unique identifiers. This overflow should be understood as a significant extension of collision lists, indicating that we have almost obtained the efficiency of the rope search instead of searching in a hash table. When the table overflows, garbage collection is only performed if the number of unused nodes has reached a threshold, e.g., 50% of all nodes; otherwise, no significant profit would be achieved. The second option in the case of overflow is to increase the size of the table of unique identifiers. Typically, the array is doubled, which provides the logarithmic complexity of the number of increases in the array size with a linear increase in the number of nodes (number of increases = $O(\log n)$). After increasing the size of the array, all elements in the array must receive a new location because one of the parameters of the hash function is the size of the array that has changed.

As previously mentioned, the computed table is commonly employed in OBDD libraries. Considering the performance aspects, the table is implemented as a hash-based cache [24]. In the hash table, in the event of a collision, a new element is added to the list. In the case of the cache, at the most k elements are stored for each hash function value, where k is a constant. If the new item must be saved, and k items already exist at the address specified by the hash function, the oldest item is replaced by the new item. This approach is the result of a trade-off between the time efficiency and the number of required memory cells.

3.2.2 Negation Attribute

Another common element in libraries is the negation attribute, which is assigned to edges (complemented or negative edges). The use of this attribute is possible due to the similarity of the OBDD representation for the function f and its negation $\bar{f}$. These diagrams differ only in the positions of the terminal nodes. For the negated function, the leaf with the value 1 is replaced with the leaf with the value 0 (Fig. 3.6). By introducing an additional edge attribute, the similarity of the diagrams can be

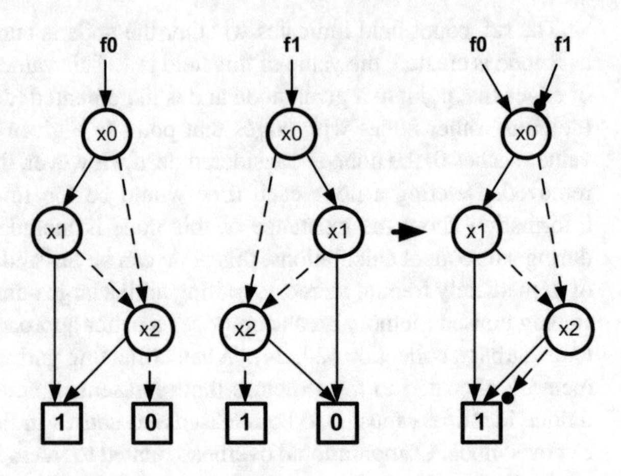

Fig. 3.6 Idea of using the edge with the negation attribute $\left(f_0 = \overline{f_1}\right)$

applied to save memory. The diagram or subdiagram indicated by the unsigned edge is interpreted as a representation of a simple function. If the attribute is assigned, the subdiagram is interpreted as a representation of the negated function. In Fig. 3.6, the negation attribute is symbolized by a black circle at the arrowhead. Using this technique, the two functions f and $\bar{f}$ can be represented by exactly the same set of nodes. In this way, you can save a significant number of memory cells. Since the indication for a terminal node with the value "0" can be replaced by an edge with the negation attribute that points to the leaf "1", in diagrams that use the said attribute, only one terminal node exist.

Using the appropriate approach, when implementing the negation attribute, the structure of Fig. 3.4 does not have to be extended to describe the diagram node. One bit is sufficient for determining the value of the negation attribute. If the addresses of the beginning of records with the given structure had only even values, then the youngest bit in the binary notation of the address would be equal to 0. The bit would not carry any useful information; thus, it could be used to store the value of the attribute. Imposing the condition on the address parity is referred to as memory alignment [25]. In addition, if the structure size is a multiple of two bytes, the alignment of the address does not cause unused memory cells between the end of the previous record and the beginning of the next record. For 32-bit processors, to accelerate memory access, a maximum alignment of 4 bytes is used; thus, several of the youngest address bits are utilized to store attribute values—including the negation attribute.

When using diagrams with edges with the negation attribute, special attention should be given to prevent loss of the canonical nature of the representation. One way to maintain canonicity is to limit the positioning of the edge with the attribute. Considering a single node with one incoming edge and two outgoing edges, we have eight possibilities to position the attribute. Using formula 3.6, we obtain four pairs of equivalent representations (Fig. 3.7).

$$\overline{\overline{x_i} \cdot f_{\overline{xi}} + x_i \cdot f_{xi}} = \overline{x_i} \cdot \overline{f_{\overline{xi}}} + x_i \cdot \overline{f_{xi}} \tag{3.6}$$

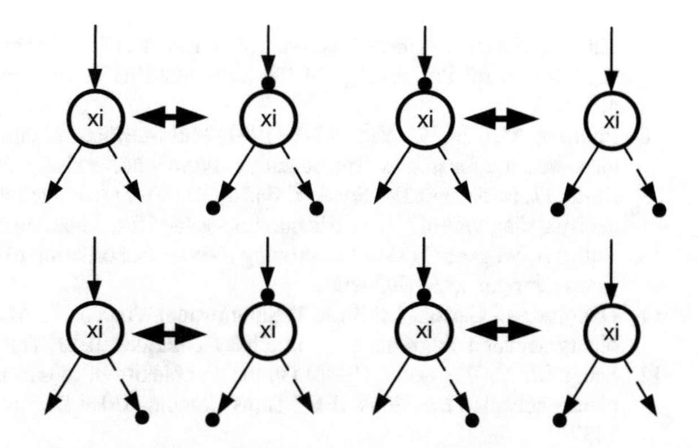

Fig. 3.7 Equivalent representations using the negation attribute

The nodes that are represented in pairs are functionally equivalent. The way to obtain the canonical character is to place a constraint on it to ensure that the outgoing edge for $x_i = 1$ cannot have the negation attribute set. In subprogram libraries, when creating a new node, the location of the attributes is selected to satisfy the previously described condition. In Fig. 3.7, this condition is illustrated by the nodes on the left in each pair.

Use of the edges with the negation attribute has several benefits:

- memory usage can theoretically be reduced by 50%,
- function negation operation can be performed at a fixed time by setting only one bit,
- binary logic operations can be significantly accelerated if one of the arguments is the negation of the other argument.

In practical applications, by using edges with the negation attribute, only 10% of memory usage can be conserved [4].

Representation of functions in the form of BDD forms the basis of many synthesis tools [26–35].

References

1. Lee CY (1959) Representation of switching circuits by binary decision programs. Bell Syst Tech J 985–999
2. Bryant RE (1986) Graph based algorithms for boolean function manipulation. IEEE Trans Comput C-35(8):677–691
3. Minato S (1996) Binary decision diagrams and applications for VLSI CAD. Kluwer Academic Publishers, Berlin
4. Meinel Ch, Theobald T (1998) Algorithms and data structures in VLSI design: OBDD—foundations and applications. Springer, Berlin
5. Minato S (1993) Zero-suppressed BDDs for set manipulation in combinatorial problems. In: 30th conference on design automation, pp 272–277
6. Drechsler R, Theobald M, Becker B (1996) Fast OFDD-based minimization of fixed polarity Reed-Muller expressions. IEEE Trans Comput 45(11):1294–1299

7. Lai Y-T, Sastry S (1992) Edge-valued binary decision diagrams for multi-level hierarchical verification. In: Proceedings of the 29th ACM/IEEE conference on design automation, pp 608–613
8. Fujita M, McGeer PC, Yang JC-Y (1997) Multi-terminal binary decision diagrams: an efficient data structure for matrix representation. Formal Methods Syst Des 10(2–3):149–169
9. Butler JT, Herscovici DS, Sasao T, Barton RJ (1997) Average and worst case number of nodes in decision diagrams of symmetric multiple-valued functions. IEEE Trans Comput 46(4):491–494
10. Bollig B, Wegener I (1996) Improving the variable ordering of OBDDs is NP-complete. IEEE Trans Comput 45(9):993–1002
11. Oliveira AL, Carloni LP, Villa T, Sangiovanni-Vincentelli AL (1998) Exact minimization of binary decision diagrams using implicit techniques. IEEE Trans Comput 47(11):1282–1296
12. Sauerhoff M, Wegener I (1996) On the complexity of minimizing the OBDD size for incompletely specified functions. IEEE Trans Comput-Aided Des Integr Circuits Syst 15(11):1435–1437
13. Yeh F-M, Kuo S-Y (1997) Variable ordering for ordered binary decision diagrams by a divide-and-conquer approach. IEE Proc Comput Digit Tech 144(5):261–266
14. Aloul F, Markov I, Sakallah K (2004) MINCE: a static global variable-ordering heuristic for SAT search and BDD manipulation. J Univ Comput Sci 10(12):1562–1596
15. Cabodi G, Camurati P, Quer S (1999) Improving the efficiency of BDD-based operators by means of partitioning. IEEE Trans Comput-Aided Des Integr Circuits Syst 18(5):545–556
16. Drechsler R, Becker B, Gockel N (1996) Genetic algorithm for variable reordering of OBDDs. Comput Digit Tech IEE Proc 143(6):364–368
17. Ishiura N, Sawada H, Yajima S (1991) Minimization of binary decision diagrams based on the exchanges of variables. In: IEEE international conference on computer aided design, pp 427–475
18. Rudell R (1993) Dynamic variable ordering for ordered decision diagram. In: IEEE international conference on computer-aided design, pp 42–47
19. Meinel C, Slobodova A (1997) Speeding up variable reordering of OBDDs. In: Computer design: VLSI in computers and processors, ICCD, pp 338–343
20. Brace KS, Rudell RL, Bryant RE (1990) Efficient implementation of a BDD package. In: 27th ACM/IEEE design automation conference, pp 40–45
21. Janssen G (2003) A consumer report on BDD packages. Integr Circuit Syst Des, pp 217–222
22. Janssen G (2001) Design of a pointerless BDD package. Workshop handouts 10th IWLS, 2001, pp 310–315
23. Aho AV, Hopcroft JE, Ullman JD (2003) Algorytmy i struktury danych. Helion
24. Long DE (1998) The design of a cache-friendly BDD library. In: International conference on computer-aided design, ICCAD, pp 639–645
25. Long DE (2008) Carnegie Mellon University BDD Library. http://wwwcs.cmu.edu/afs/cs/project/modck/pub/www/bdd/html
26. Cheng L, Chen D, Wong MDF (2007) DDBDD: Delay-driven BDD synthesis for FPGAs. In: Design automation conference, 2007, DAC '07. 44th ACM/IEEE, pp 910–915
27. Kubica M, Kania D (2016) SMTBDD: new form of BDD for logic synthesis. Int J Electr Telecommun 62(1):33–41
28. Kubica M, Kania D (2017) Area-oriented technology mapping for LUT-based logic blocks. Int J Appl Math Comput Sci 27(1):207–222
29. Kubica M, Kania D (2017) Decomposition of multi-output functions oriented to configurability of logic blocks. Bull Polish Acad Sci Tech Sci 65(3):317–331
30. Kubica M, Opara A, Kania D (2017) Logic synthesis for FPGAs based on cutting of BDD. Microprocess Microsyst 52:173–187
31. Kubica M, Kania D (2019) Technology mapping oriented to adaptive logic modules. Bull Polish Acad Sci Tech Sci 67(5):947–956
32. Opara A, Kubica M, Kania D (2018) Strategy of logic synthesis using MTBDD dedicated to FPGA Integration. VLSI J 62:142–158

33. Opara A, Kubica M, Kania D (2019) Methods of improving time efficiency of decomposition dedicated at FPGA structures and using BDD in the process of cyber-physical synthesis. IEEE Access 7:20619–20631
34. Vemuri N, Kalla P, Tessier R (2002) A BDD-based logic synthesis for LUT-based FPGAs. ACM Trans Des Autom Electron Devices 7(4):501–525
35. Yang C, Ciesielski M (2002) BDS: a BDD-based logic optimization system. IEEE Trans Comput Aided Des Integr Circuits Syst 21(7):866–876

Chapter 4
Theoretical Basis of Decomposition

Function decomposition is a key elements of logic synthesis. A substantial interest in decomposition has been observed since the introduction of FPGAs. Decomposition algorithms exist in the literature for PLA matrices and CPLDs [1–13]. In this case, however, the importance of decomposition for optimizing the designed system is apparent. In the case of FPGAs, decomposition is a key element of the synthesis, which constitutes a mathematical model of the division of the implemented system among the logic blocks contained inside the FPGA. In the simplest case, these blocks are LUTs with a fixed number of k inputs. This chapter provides a theoretical basis for decomposing a function focused on FPGA.

4.1 Functional Decomposition Theorem

The classical decomposition theory is based on the decomposition theorem presented by Ashenhurst [14] and developed by Curtis [15]. This theorem will be discussed in a general form, i.e., focused on the multioutput function with n-inputs and m-outputs.

Let *f* be the n-input and m-output logic function, i.e., $f : B^n \rightarrow B^m$, where $B = \{0,1\}$. Let $I = \{I_{n-1}, \ldots, I_1, I_0\}$, $Y = \{Y_{m-1}, \ldots, Y_1, Y_0\}$ be a set of input variables and output variables, respectively.

A function $f : B^n \rightarrow B^m$ is subject to simple serial decomposition, i.e.,

$$F(X_b, X_f) = F(g_1(X_b), g_2(X_b), \ldots, g_p(X_p), X_f) \tag{4.1}$$

if and only if the column multiplicity of the partition matrix (Karnaugh map) $\nu(X_b|\, X_f) \leq 2^p$. The *Xb* and *Xf* sets are referred to as the bound set and the free set, respectively.

where $X_b \cup X_f = \{I_{n-1}, \ldots, I_1, I_0\}$ and $X_b \cap X_f = \phi$.

Applying this decomposition model causes the partitioning of a multioutput block that is characterized by an ordered pair of numbers (number of inputs, number of

M. Kubica et al., *Technology Mapping for LUT-Based FPGA*, Lecture Notes in Electrical Engineering 713, https://doi.org/10.1007/978-3-030-60488-2_4

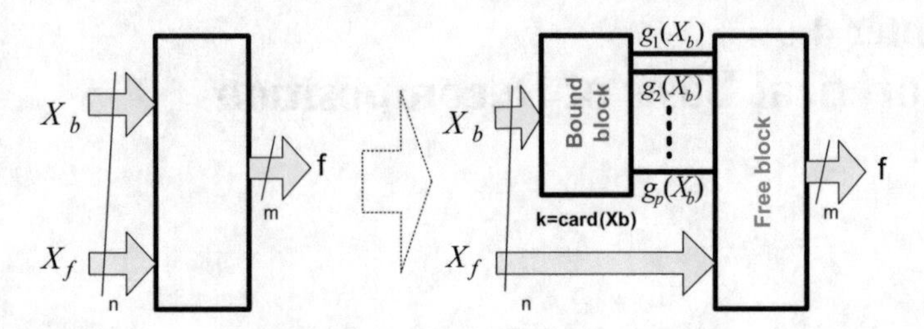

Fig. 4.1 Simple serial decomposition

outputs) = (*n*, *m*) into two new blocks described by parameters (*k*, *p*) and (*n* − *k* + *p*, *m*), where *k* = *card*(*Xb*). Partitioning of the circuit, which corresponds to decomposition of a multioutput function, is presented in Fig. 4.1.[1]

The model of decomposition, which is illustrated in Fig. 4.1, is referred to as a simple serial decomposition. This kind of decomposition includes a single bound block and a single free block. The connections between a bound block and a free block correspond to the bound functions $g_1(X_b), \ldots, g_p(X_b)$. Minimization of the number of bound functions (*p*) is particularly essential from the point of view of the efficiency of technology mapping into an LUT-based FPGA (see Footnote 1).

An example of a function subject to simple serial decomposition is shown in Fig. 4.2.

The function of the 5 variables $\{x_0, x_1, x_2, x_3$, and $x_4\}$ in Fig. 4.2a was described using Karnaugh map. In the Karnaugh map, two column patterns can be distinguished with the symbols A and B. The column multiplicity of this map is 2. To distinguish among the column patterns, only a single bit, which is associated with the a single ($p = 1$) bound function g, is needed. After the appropriate codes are assigned to the patterns A and B, they can be assigned the bound function, as shown in Fig. 4.2b. After entering the bound function, the function description from Fig. 4.2a can be presented in the form from Fig. 4.2c. This function is implemented in a free block and is referred to as a free function. The resulting division is shown in Fig. 4.2d.

Due to a substantial number of variables, more complex models of decomposition, such as multiple and iterative decomposition, are needed [15, 17]. In the case of complex models of decomposition, choosing an appropriate decomposition path first and then gradually limiting the number of variables until the number of inputs of subcircuits is lower or equal to k is important.

[1]

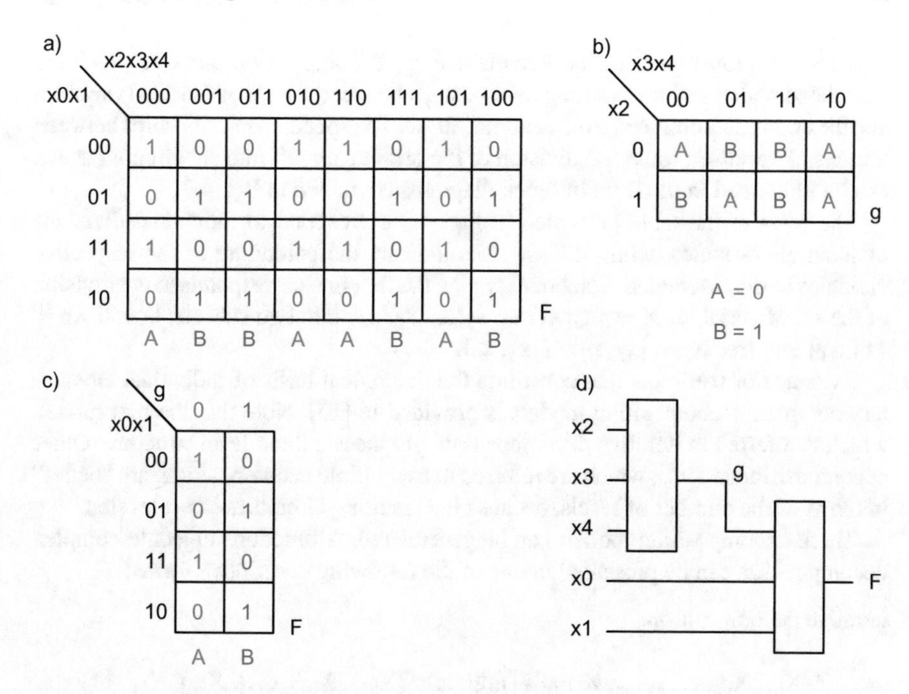

Fig. 4.2 Idea of simple serial decomposition

4.2 Complex Decomposition Models

The process of function decomposition (circuit partition) is usually a multistage process. A single step is related to the search for the most advantageous partition of function variables [18, 19]. By analyzing the various breakdowns of function variables, the obtained information can be used to obtain the appropriate complex decomposition. In general, several basic complex decomposition models can be distinguished directly from the relationship between two variables of bounded sets [15, 20].

Let the function f(X) be subject to two different decompositions: $f(X) = f(X_{b1}, X_{f1})$ and $f(X) = f(X_{b2}, X_{f2})$, in which Xb1, Xb2 are bound sets of individual divisions of the set of arguments X. Different relationships may exist between the bound sets, e.g., $X_{b1} \supset X_{b2}$ or $X_{b1} \cap X_{b2} = \emptyset$. In the general case, five different situations that correspond to the basic model of complex decomposition can occur [15, 21].

1. $X_{b1} \supset X_{b2}$
2. $X_{b1} \cap X_{b2} = \emptyset$ and $X_{b1} \cup X_{b2} = X$
3. $X_{b1} \cap X_{b2} = \emptyset$ and $X_{b1} \cup X_{b2} \neq X$
4. $X_{b1} \cap X_{b2} \neq \emptyset$ and $X_{b1} \cup X_{b2} = X$
5. $X_{b1} \cap X_{b2} \neq \emptyset$ and $X_{b1} \cup X_{b2} \neq X$.

These relationships describe five elementary decomposition models, which are associated with the corresponding relationships between the set of function variables and the bound sets obtained from decomposition. The specified relationships between the sets of variables cause the division of the original circuit into subcircuits, which can be illustrated in the form of block diagrams contained in Fig. 4.3.

The rows in the table presented in Fig. 4.3 correspond to individual divisions of function variables, while the asterisks indicate the belonging of the respective variables to the associated sets. For example, the first line corresponds to the division of the set of variables $X = \{x_0, x_1, x_2, x_3, x_4, x_5, x_6\}$ into two subsets: bound $Xb = \{x_0, x_1\}$ and free $X_f = \{x_2, x_3, x_4, x_5, x_6\}$.

Evidence of theorems that constitute the theoretical basis of individual elementary complex decomposition models is provided in [15]. Note that the first model, which is referred to iterative decomposition, produces a three-level structure. Other decomposition models, which are referred to as multiple decomposition, are "better" in terms of the number of levels, because the resulting circuits are two-leveled.

These decomposition models can be generalized. A function subject to complex decomposition can be presented in one of the following descriptive forms:

iterative decomposition:

$$f(X_{b1}, X_{b2}, \ldots, X_{bn}, X_f) = F[\eta[\xi[\ldots \varphi[(X_{b1}), X_{b2}], \ldots], X_{bn}], X_f] \quad \text{or}$$

multiple decomposition:

$$f(X_{b1}, X_{b2}, \ldots, X_{bn}, X_f) = F[\chi(X_{b1}), \xi(X_{b2}), \ldots, \varphi(X_{bn}), X_f].$$

In addition, the presented models can be generalized to cases in which decompositions relate to sets of functions or are obtained from divisions for which the column multiplicity is greater than two. A certain problem is the way to determine the functions implemented in individual bound sets and free blocks created after applying the appropriate complex decomposition.

4.3 Iterative Decomposition

Iterative decomposition is a frequently employed model of complex decomposition. Iterative decomposition of a function becomes possible when at least two simple serial decompositions, for which bound sets satisfy the condition $X_{b1} \supset X_{b2}$, can be obtained in the set of argument breakdowns.

A formal description of iterative decomposition of a multioutput function can be presented in the form of an iterative decomposition theorem.

Theorem 4.1 (Iterative decomposition)

The function $f: B^n \rightarrow B^m$ is subject to q various decompositions, that is,

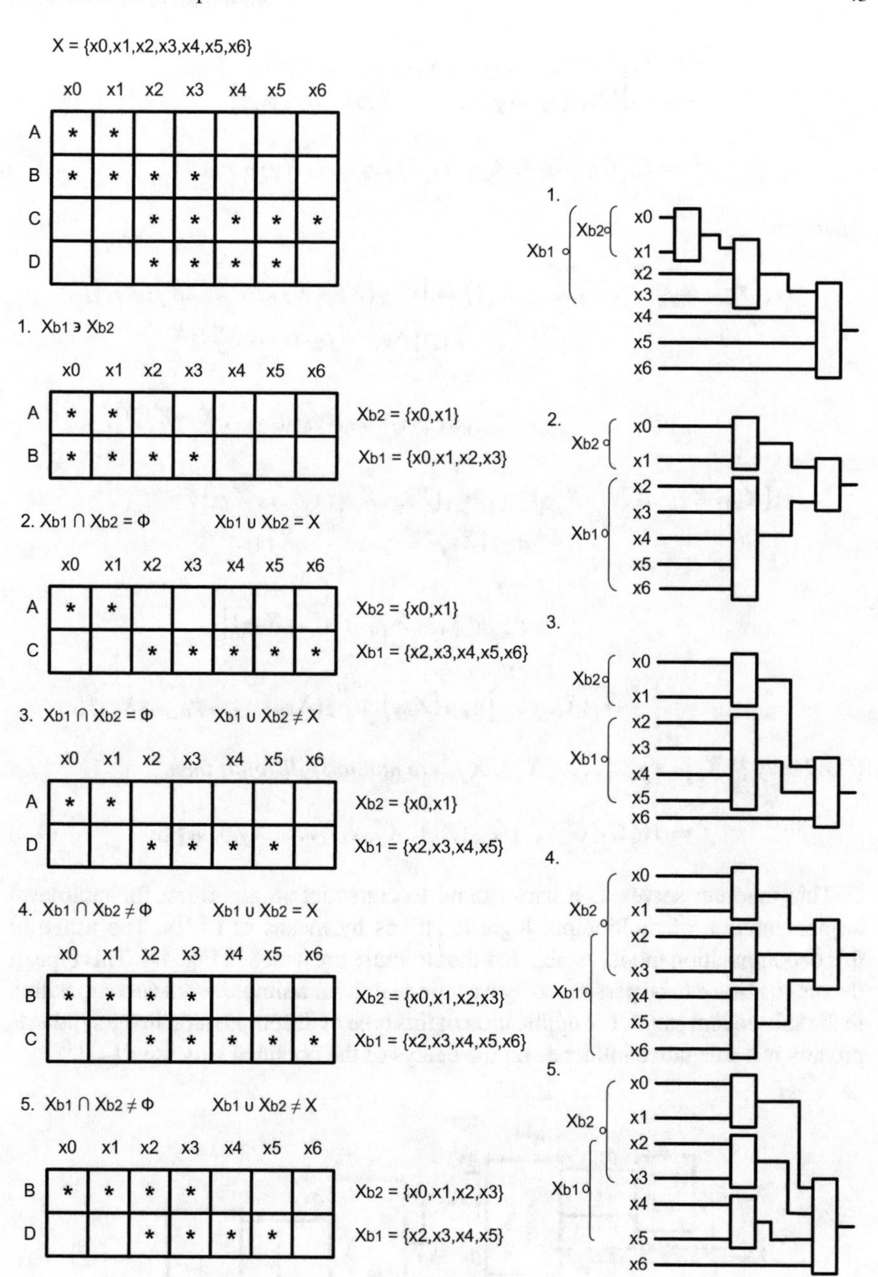

X = {x0,x1,x2,x3,x4,x5,x6}
x0 x1 x2 x3 x4 x5 x6
A
B
C
D
1. Xb1 ∋ Xb2
Xb2 = {x0,x1}
Xb1 = {x0,x1,x2,x3}
2. Xb1 ∩ Xb2 = Φ Xb1 ∪ Xb2 = X
Xb2 = {x0,x1}
Xb1 = {x2,x3,x4,x5,x6}
3. Xb1 ∩ Xb2 = Φ Xb1 ∪ Xb2 ≠ X
Xb2 = {x0,x1}
Xb1 = {x2,x3,x4,x5}
4. Xb1 ∩ Xb2 ≠ Φ Xb1 ∪ Xb2 = X
Xb2 = {x0,x1,x2,x3}
Xb1 = {x2,x3,x4,x5,x6}
5. Xb1 ∩ Xb2 ≠ Φ Xb1 ∪ Xb2 ≠ X
Xb2 = {x0,x1,x2,x3}
Xb1 = {x2,x3,x4,x5}
Xb1
Xb2

Fig. 4.3 Complex decompositions of an example function

$$f = F_1\left[G_1\left(X_{bq}, X_{fq-1}, X_{fq-2}, \ldots, X_{f1}\right), \quad X_f\right],$$
$$f = F_2\left[G_2\left(X_{bq}, X_{fq-1}, \ldots, X_{f2}\right) \quad X_{f1}X_f,\right]$$
$$\ldots$$
$$f = F_q\left[Gq\left(X_{bq}\right), X_{fq-1}, X_{fq-2}, \ldots, X_{f2}X_{f1}X_f\right],$$

where

$$\begin{aligned} G_1\left(X_{bq}, X_{fq-1}, \ldots, X_{f2}, X_{f1}\right) = \big[&g_{1\_1}\left(X_{bq}, X_{fq-1}, \ldots, X_{f2}X_{f1}\right), \\ &g_{1\_2}\left(X_{bq}, X_{fq-1}, \ldots, X_{f2}X_{f_1}\right), \\ &\ldots, \\ &g_{1\_p1}\left(X_{bq}, X_{fq-1}, \ldots, X_{f2}X_{f1}\right)\big], \end{aligned}$$

$$\begin{aligned} G_2\left(X_{bq}, X_{fq-1}, \ldots, X_{f2}\right) = \big[&g_{2\_1}\left(X_{bq}, X_{fq-1}, \ldots, X_{f2}\right), \\ &g_{2\_2}\left(X_{bq}, X_{fq-1}, \ldots, X_{f2}\right), \\ &\ldots \\ &g_{2\_p2}\left(X_{bq}, X_{fq-1}, \ldots, X_{f2}\right)\big] \\ &\ldots \end{aligned}$$

$$G_q\left(X_{bq}\right) = \left[g_{q\_1}\left(X_{bq}\right), g_{q\_2}\left(X_{bq}\right), \ldots, g_{q\_pq}\left(X_{bq}\right)\right],$$

if and only if $X_{bq}, X_{fq-1}, \ldots, X_{f1}, X_f$, *are mutually disjoint; then*

$$f = F_0[G_1[G_2 \ldots [G_{bq}(X_q), X_{fq-1}], \ldots, X_{f1}), X_f]. \tag{4.2}$$

This theorem serves as a background to construct an algorithm for multilevel implementation of multioutput logic functions by means of LUTs. The usage of this decomposition model establishes the structure presented in Fig. 4.4. This type of decomposition can be performed by a cyclic search for a simple serial decomposition in the subsequent steps. The application of this type of decomposition in the synthesis process has a negative influence on the delays of the obtained structures [22].

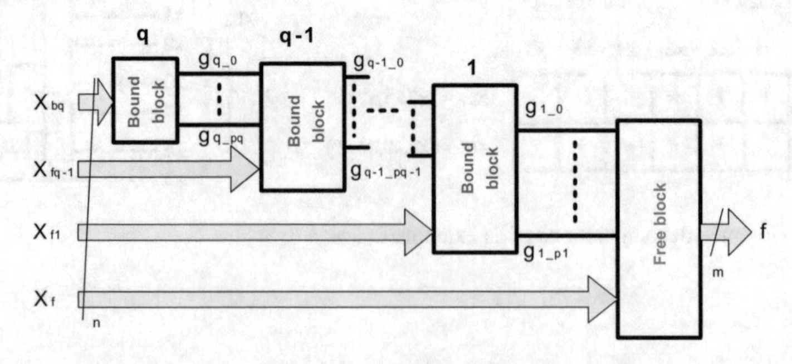

Fig. 4.4 Structure of the circuit after the usage of iterative decomposition

4.4 Multiple Decomposition

The second model of complex decomposition is multiple decomposition. Assuming the separation of individual bound sets, the formal description of multiple decomposition of a multioutput function can be presented in the form of the multiple decomposition theorem.

Theorem 4.2 (Multiple decomposition)

A function $f\colon B^n \to B^m$ is subject to q different decompositions:

$$\begin{aligned} f &= F_1\big[G_1(X_{b1}),\quad X_{b2}X_{b3},\ldots,X_{bq},\quad X_f\big],\\ f &= F_2\big[X_{b1},G_2(X_{b2}),\quad X_{b3},\ldots,X_{bq},X_f\big],\\ &\ldots\\ f &= F_q\big[X_{b1},X_{b2},X_{b3},\ldots,\quad G_q\big(X_{bq}\big),\quad X_f\big], \end{aligned}$$

where

$$\begin{aligned} G_1(X_{b1}) &= \big[g_{1\_1}(X_{b1}),g_{1\_2}(X_{b1}),\ldots,g_{1\_p1}(X_{b1})\big],\\ G_2(X_{b2}) &= \big[g_{2\_1}(X_{b2}),g_{2\_2}(X_{b2}),\ldots,g_{2\_p2}(X_{b2})\big],\\ &\ldots\\ G_q\big(X_{bq}\big) &= \big[g_{q\_1}\big(X_{bq}\big),g_{q\_2}\big(X_{bq}\big),\ldots,g_{q\_pq}\big(X_{bq}\big)\big], \end{aligned}$$

if and only if $X_{bq}, X_{bq-1}, \ldots, X_{b1}$, Xf are mutually disjoint; then

$$f = F\big[G_1(X_{b1}),G_2(X_{b2}),\ldots,G_q\big(X_{bq}\big),X_f\big] \tag{4.3}$$

The use of this theorem produces the circuits partition presented in Fig. 4.5 [22]. Multiple decomposition does not significantly increase the number of logic levels in contrast to iterative decomposition.

We attempt to obtain complex decompositions of the function $f\colon B^5 \to B$, which is described using the Karnaugh grid shown in Fig. 4.6a. Assuming that the cases that enable the limitation of the number of bound and free block inputs are interesting, a set of decompositions has been obtained, as symbolically shown in the table in Fig. 4.6b. The individual columns contain the number of decompositions, a symbolic description of the division of variables, the numbers that specify the column multiplicity of the considered division, and the number of bound functions. For example, line 6 relates to the case in which the bound set and free set contain $X_b = \{x_2, x_3, x_4\}$ and $X_f = \{x_0, x_1\}$, respectively. For this division of variables, a Karnaugh column multiplicity of 2 (two types of column patterns in the Karnaugh map in Fig. 4.6a) was obtained.

The process of selecting the appropriate complex decomposition usually directly depends on the parameters of the logic blocks in the implementation process. The

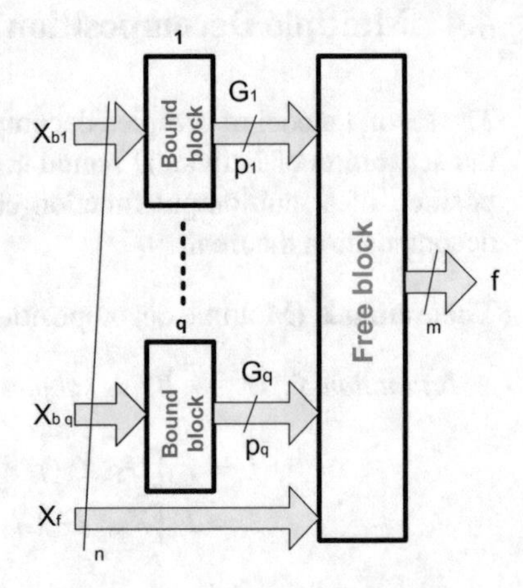

Fig. 4.5 Structure of the circuit after using multiple decomposition

number of inputs of LUT is decisive. We assume that we want to implement the analyzed function on LUT blocks with 3 inputs (LUT3/1). This implementation is made possible by multiple decomposition, which is obtained after noticing the two simple serial decompositions represented in the table in the rows with the indexes 6 and 13. The result of multiple decomposition yields the solution shown in Fig. 4.6c. The Karnaugh map contains two kinds of columns and two types of rows, enabling the use of the multiple decomposition model implemented using two LUT3/1 blocks and one LUT2/1 block.

Many synthesis strategies performed in the LUT-based FPGAs use the elements of iterative or multiple decomposition in a direct or indirect way [16, 22–26]. In subsequent decomposition steps, individual decomposition models can be interchangeably utilized, depending on which solution at a given step is better. This situation can be described as mixed decomposition. In general, all decomposition models discussed in this chapter belong to the serial decomposition group. In addition to the various serial decomposition models, special attention should be given to parallel decomposition [27–29], which is sometimes referred to as single-level decomposition [30] and is related to the decomposition model, whose essence is the division of the multi-output function. A skillful search for insignificant variables for individual single-output functions and the use of effective algorithms for reducing variables [31–33] is particularly important in this type of decomposition model.

The use of both serial and parallel decomposition in the decomposition process generates a multilevel structure, in which the division of inputs or outputs is performed in subsequent stages. This type of LUT-based network creation is sometimes associated with the concept of balanced decomposition [34–36]. Numerous theoretical considerations related to the problem of coding and aimed at minimizing the number of free block inputs are provided in [37–40].

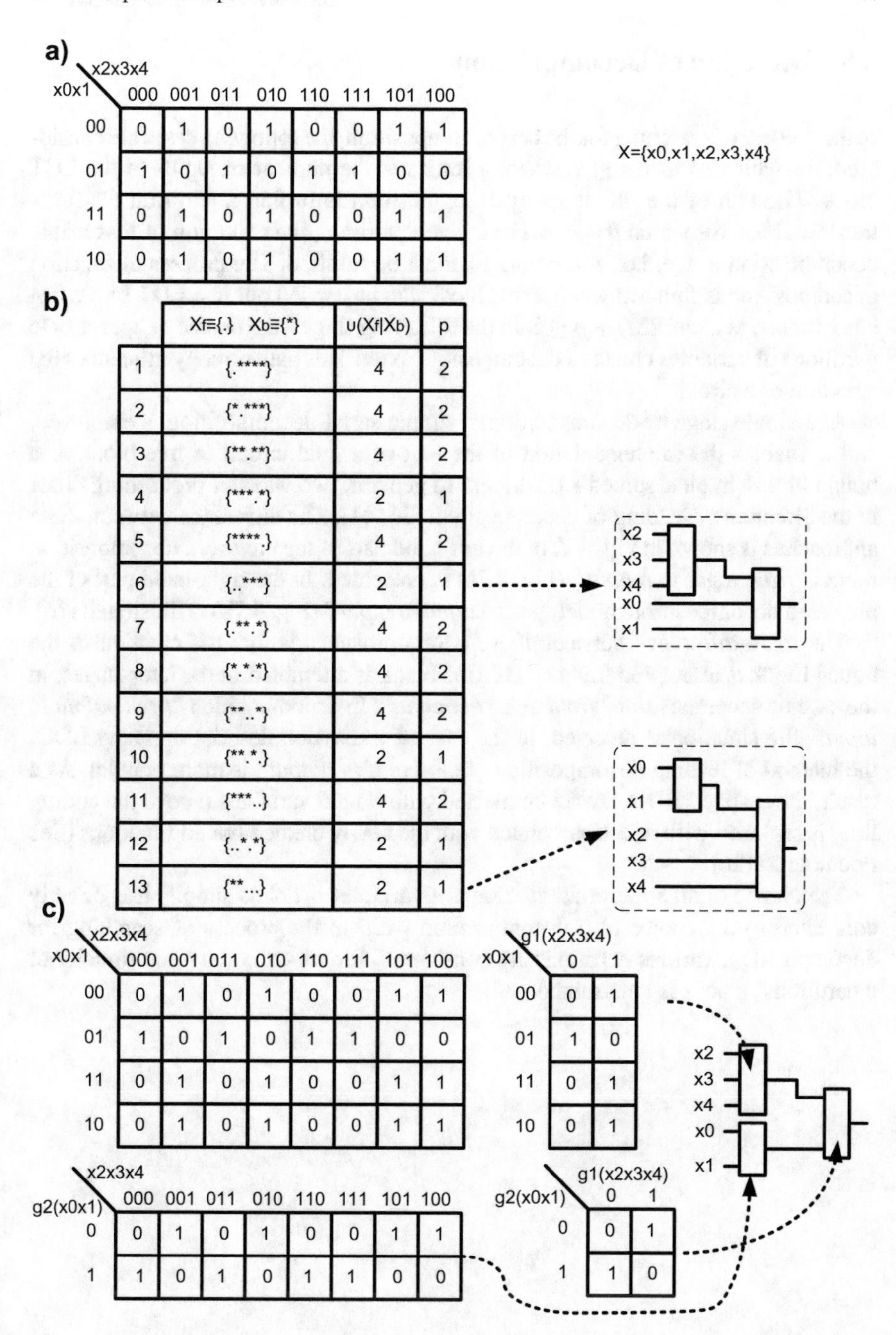

Fig. 4.6 Example of multiple decomposition

4.5 Direction of Decomposition

In the process of searching for the best decomposition, the following cases are considered, for which *card(Xb)* $\leq k$, where k indicates the number of inputs of the LUT block. The aim of the i-th stage of decomposition is to find a partition for function variables, for which $p = min$. For a given partition, the condition of reasonable decomposition usage, i.e., $p <$ *card(Xb)*, must be fulfilled. The process of function decomposition is finished when a free block can be carried out in a LUT block that has k inputs, i.e., *card(Xf)* $+ p \leq k$. In the following stages, the choice of appropriate partitions of variables creates a decomposition path. This path directly influences the structure of a circuit.[2]

At a single stage of decomposition, a simple serial decomposition is employed, and it enables the implementation of the part of a final circuit (a free block or a bound block) in an assumed LUT block. In general, two ways of proceeding differ in the direction of leading of decomposition [26, 41]. The difference between these approaches is shown in Fig. 4.7. In the left-hand part of the picture, a decomposition model *"from input to output"* (Fig. 4.7a) is presented. In the right-hand part of the picture, a decomposition model *"from output to input"* (Fig. 4.7b) is illustrated [26].

The main difference between these two approaches is that the creation of the bound block is attempted first and the free block is attempted in the later stages, in the case of decomposition "*from input to output*". In decomposition "*from output to input*", the situation is reversed. In the case of a function description using BDD, the method of leading decomposition "*from input to output*" is more popular. As a result, the mapped BDD extract connected with bound variables (above the cutting line) is replaced with nodes associated with the newly created bound functions (See Footnote 2) [26].

The choice of an appropriate partition of variables is not as simple; it is directly connected to the choice of a decomposition path. In the process of searching for decomposition, various partitions are considered, for which column multiplicity of a partitions' matrix is determined.

---

[2] © Reprinted from Integration: Opara et al. [26], Copyright (2020), with permission from Elsevier.

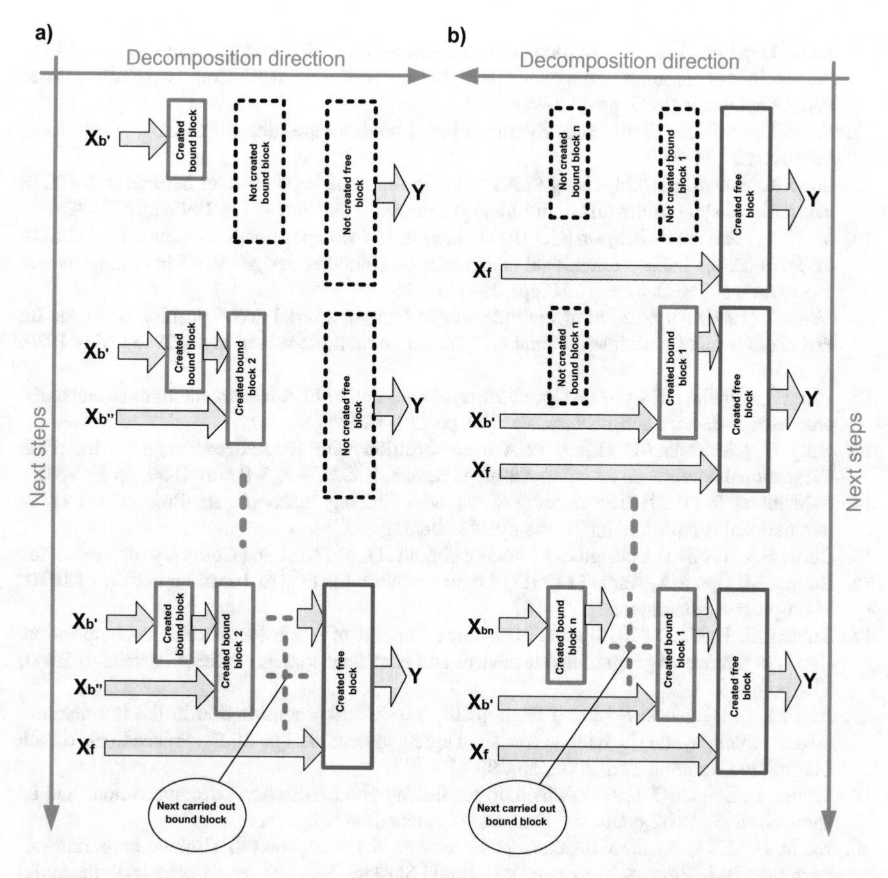

Fig. 4.7 Direction of leading of decomposition: **a** "*from input to output*" and **b** "*from output to input* (see Footnote 2)

References

1. Ciesielski MJ, Yang S (1992) PLADE: A two-stage PLA decomposition. IEEE Trans Comput-Aided Des 11(8):943 954
2. Devadas S, Wang AR, Newton AR, Sangiovanni-Vincentelli A, Boolean decomposition of programmable logic arrays. In: IEEE custom integrated circuits conference, May 1988, pp 2.5.1–2.5.5
3. Devadas S, Wang AR, Newton AR, Sangiovanni-Vincentelli A (1988) Boolean decomposition in multi-level logic optimization. In: Digest of technical papers, IEEE international conference on computer-aided design, ICCAD-88, 7–10 Nov 1988, pp 290–293
4. Kania D (2015) Logic decomposition for PAL-based CPLDs. J Circuit Syst Comput 24(3):1–27
5. Kania D, Kulisz J, Milik A (2005) A novel method of two-stage decomposition dedicated for PAL-based CPLDs. In: Proceedings of Euromicro symposium on digital system design, IEEE Computer Society Press, Porto, September, 2005, pp 114–121
6. Kania D, Kulisz J (2007) Logic synthesis for PAL-based CPLD-s based on two-stage decomposition. J Syst Softw 80:1129–1141

7. Kania D, Milik A, Kulisz J (2005) Decomposition of multiple-output functions for CPLDs, Proceedings of euromicro symposium on digital system design. IEEE Computer Society Press, Porto, September, 2005, pp 442–449
8. Kania D, Milik A (2010) Logic Synthesis based on decomposition for CPLDs. Microproces Microsyst 34:25–38
9. Lee KK, Wong DF (1998) Using PLAs to design universal logic modules in FPGAs. In: IEEE international symposium on circuits and systems, vol 6, 31 May–3 Jun 1998, pp 421–425
10. Malik A, Harrison D, Brayton RK (1991) Three-level decomposition with application to PLDs. In: Proceedings IEEE international conference on computer design: VLSI in computers and processors, 14–16 October 1991, pp 628–633
11. Sasao T (1989) Application of multiple-valued logic to a serial decomposition of PLAs. In: Proceedings nineteenth international symposium on multiple-valued logic, 29–31 May 1989, pp 264–271
12. Wang L, Almaini AEA (2002) Optimisation of reed-muller PLA implementations circuits. IEE proceedings-devices and systems, vol 149, pp 119–128
13. Yang C, Ciesielski MJ (1989) PLA decomposition with generalized decoders. In: IEEE international conference on computer-aided design, ICCAD-89, 5–9 Nov 1989, pp 312–315
14. Ashenhurst R (1957) The decomposition od switching functions. In: Proceedings of an international symposium on the theory of switching
15. Curtis HA (1962) The design of switching circuits. D. van Nostrand Company Inc., Princeton
16. Kubica M, Opara A, Kania D (2017) Logic synthesis for FPGAs based on cutting of BDD. Microprocess Microsyst 52:173–187
17. Kubica M, Kania D (2015) SMTBDD : new concept of graph for function decomposition, IFAC conference on programmable devices and embedded systems PDeS, Cracow 2015, VII, 519, pp 61–66
18. Jozwiak L, Biegahski S (2005) High-quality sub-function construction in the information-driven circuit synthesis with gates. In: Digital system design 2005. Proceedings of 8th Euromicro Conference on, 2005, pp 450–459
19. Sechen C, Stanion T (1995) A method for finding good Ashenhurst decompositions and its application to FPGA synthesis, 32nd design automation conference
20. Kania D (2004) Synteza logiczna przeznaczona dla matrycowych struktur programowalnych typu PAL, Zeszyty Naukowe Politechniki Śląskiej, Nr 1619. Wydawnictwo Politechniki Śląskiej, Gliwice
21. Kania D (2012) Układy Logiki Programowalnej - Podstawy syntezy i sposoby odwzorowania technologicznego. PWN, Warszawa
22. Kubica M, Kania D (2017) Area-oriented technology mapping for LUT-based logic blocks. Int J Appl Math Comput Sci 27(1):207–222
23. Kubica M, Kania D, Opara A (2016) Decomposition time effectiveness for various synthesis strategies dedicated to FPGA structures. In: International conference of computational methods in sciences and engineering, American Instytut of Physics, Athens, 17 Mar 2016, Seria: AIP conference proceedings; vol 1790
24. Kubica M, Kania D (2016) SMTBDD : new form of BDD for logic synthesis. Int J Electr Telecommun 62(1):33–41
25. Kubica M, Kania D (2017) Decomposition of multi-output functions oriented to configurability of logic blocks. Bull Polish Acad Sci Tech Sci 65(3):317–331
26. Opara A, Kubica M, Kania D (2018) Strategy of logic synthesis using MTBDD dedicated to FPGA. Integr VLSI J 62:142–158
27. Luba T (2000) Synteza układów logicznych. Wyższa Szkoła Informatyki Stosowanej i Zarządzania, Warszawa
28. Luba T (red.) (2003) Rawski M, Tomaszewicz P, Zwierzchowski B, Synteza układów cyfrowych, WKŁ Warszawa
29. Luba T, Nowicka M, Perkowski M, Rawski M (1998) Nowoczesna synteza logiczna. Oficyna Wydawnicza Politechniki Warszawskiej, Warszawa

30. Fiser P, Kubátová H (2005) Output grouping-based decomposition of logic functions. In: Proceedings of 8th IEEE design and diagnostics of electronic circuits and systems workshop (DDECS'05), Sopron, Hungary, 13 16.4.2005, pp 137–144
31. Mazurkiewicz T, Łuba T (2019) Non-disjoint decomposition using r-admissibility and graph coloring and its application in index generation functions minimization. In: Mixed design of integrated circuits and systems 2019 MIXDES—26th international conference, pp 252–256
32. Rawski M (2011) Application of indexed partition calculus in logic synthesis of Boolean functions for FPGAs.Int J Electr Telecommun 57(2):209–216
33. Sasao T, Matsuura K, Iguchi Y (2017) An algorithm to find optimum support-reducing decompositions for index generation functions. In: Design, automation & test in Europe conference & exhibition (DATE), Lausanne, Switzerland, 17–31 March 2017
34. Brzozowski J, Luba T (2003) Decomposition of Boolean functions specified by cubes. J Multiple-Valued Logic Soft Comput 9:377–417
35. Jozwiak L, Chojnacki A (2003) Effective and efficient FPGA synthesis through general functional decomposition. J Syst Architect 49(4–6):247–265
36. Wisniewski M, Deniziak S (2018) BMB synthesis of binary functions using symbolic functional decomposition for LUT-based FPGAs. J. Parallel Distrib Comput 120 May 2018
37. Lai Y, Pan K, Pedram M (1994) FPGA synthesis using function decomposition. In: Proceedings of the IEEE international conference on computer design. Cambridge, pp 30–35
38. Scholl C (2001) Functional decomposition with application to FPGA synthesis. Kluwer Academic Publishers, Boston
39. Scholl C, Molitor P (1994) Efficient ROBDD based computation of common decomposition functions of multi-output Boolean functions. IFIP Workshop on Logic and Architecture Synthesis, Grenoble, pp 61–70
40. Wurth B, Schlichtmann U, Eckl K, Antreich K (1999) Functional multiple-output decomposition with application to technology mapping for lookup table-based FPGAs. ACM Trans Design Autom Electron Syst 4(3):313–350
41. Hrynkiewicz E, Kania D (2003) Impact of decomposition direction on synthesis effectiveness. In: Programmable devices and systems, PDS'03, February 11–13, Ostrava, pp 144–149, 2003

Chapter 5
Decomposition of Functions Described Using BDD

Representation of functions in the form of BDD forces the necessity of introducing appropriate algorithms of decomposition [1–12]. All previously described decomposition models can be realized by making the appropriate horizontal cuts of the BDD diagram. To obtain the correct solution, the appropriate decomposition rules, to which this chapter is devoted, must be observed. In general, decomposition methods for representing functions in the form of BDD can be divided into methods that use a single cut line and methods that use multiple cut lines. First, methods that use a single cutting line will be discussed. The authors that subsequently write the BDD by default assume that the BDD is a reduced and orderly form of BDD (ROBDD).

5.1 Methods for Performing Function Decomposition Using Single Cuts of the BDD

Simple serial decomposition is the most straightforward decomposition model and is the basis of complex decomposition models, such as iterative and multiple decomposition [13, 14].

5.1.1 Simple Serial Decomposition—Single Cut Method

For function representation in BDD, the process of performing a simple serial decomposition relies on a horizontal cut of a diagram [15]. The part above the cutting line is associated with the bound set *Xb*, and the part below the cutting line is connected with the free set *Xf*. The number (p) of bound functions (i.e., the number of connections between a bound block and a free block) corresponds to the number of bits that are needed to differentiate the nodes; these nodes are the nodes from the part of the diagram below the cutting line. These nodes are pointed by edges from the

M. Kubica et al., *Technology Mapping for LUT-Based FPGA*, Lecture Notes in Electrical Engineering 713, https://doi.org/10.1007/978-3-030-60488-2_5

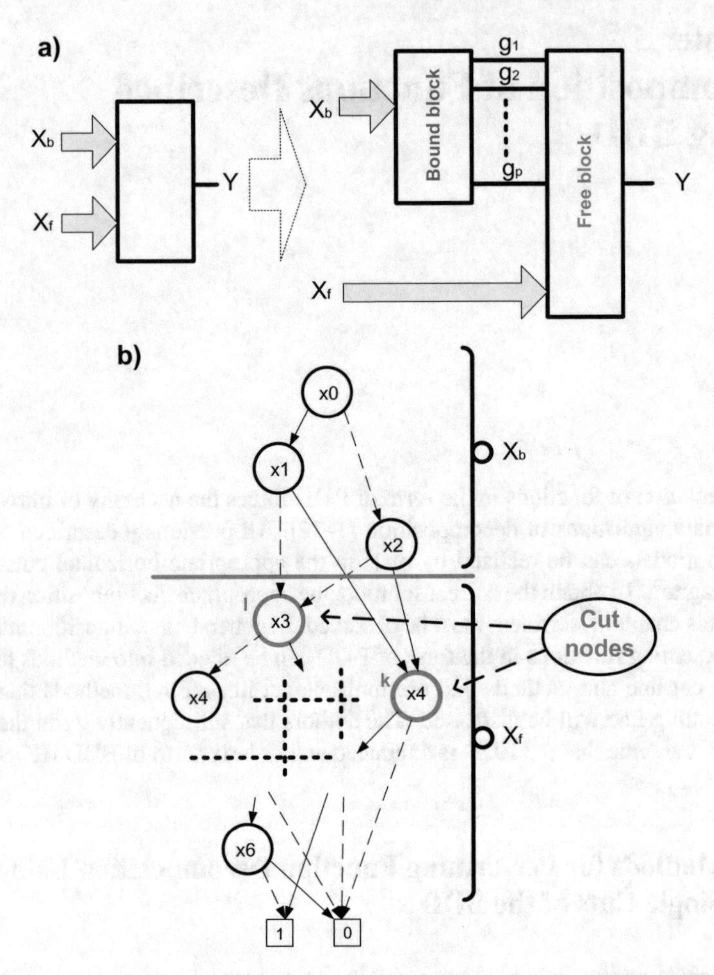

Fig. 5.1 A simple serial decomposition of function, **a** the result of decomposition, and **b** BDD

nodes placed above the cutting line. Because the number of cut nodes is equal to the column multiplicity of a partition matrix, the number of essential bound functions is (5.1).[1]

$$p = lg_2\lceil \nu(X_f|X_b)\rceil \quad (5.1)$$

In the case of BDD, the symbol ν represents the number of cut nodes. Both sets do not contain common variables ($Xb \cap Xf = \Phi$). An illustration of simple serial decomposition from the cutting of a BDD diagram is presented in Fig. 5.1.

[1]© Reprinted from Kubica et al. [5], Copyright (2020), with permission from Elsevier.

A simple serial decomposition is defined as a decomposition that corresponds to one bound and one free block [13]. In general, the number of inputs of a combinational circuit is sufficient, and the model is too small.

5.1.2 *Iterative Decomposition—Single Cut Method*

The implementation of iterative decomposition is the sequential execution of subsequent decomposition steps. The appropriate cutting line is entered in each step, and then the upper part of the diagram is replaced with a fragment that contains vertices that represent the newly created bound functions. Each decomposition step produces the next logic level in the created LUT-based network. The concept of searching for iterative decomposition using BDD is illustrated in Fig. 5.2.

A function represented by a BDD diagram (Fig. 5.2a) is subjected to simple serial decomposition by cutting the diagram. This cutting divides the variables into two subsets: $X_b = \{x_0, x_1, x_2, x_3\}$ and $X_f = \{x_4, x_5, x_6, x_7, x_8\}$. The top section, which describes a bound function performed by a bound block (1), is replaced with nodes that are connected with the single bound function *g0*. As a result of this

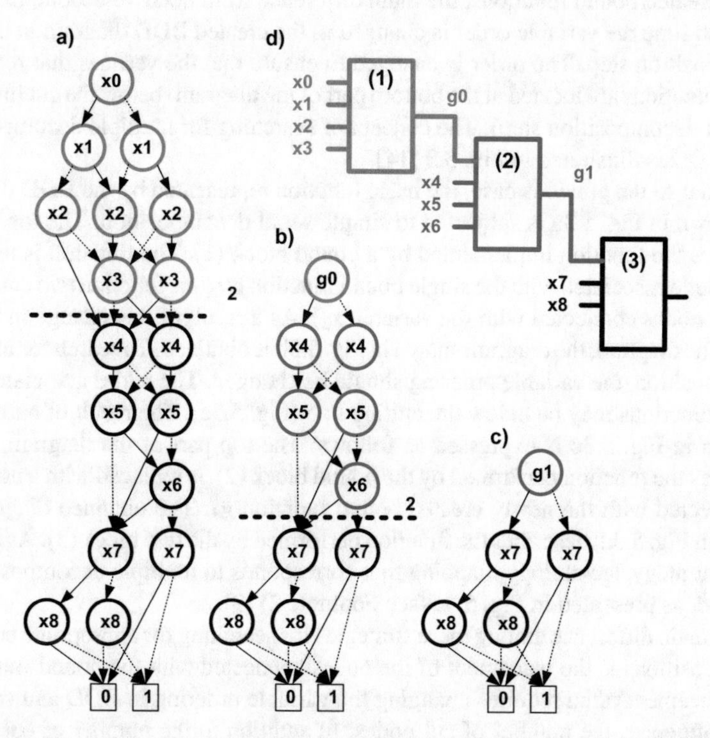

Fig. 5.2 Iterative decomposition: **a–c** show the initial steps; **d**) shows the implementation of function by means of LUTs (see Footnote 2) [9]

decomposition, a single bound function is created, since two cut nodes associated with the variable x_4 were created in the process of BDD cutting. After the first step of decomposition, a diagram is obtained (Fig. 5.2b) and is subjected to the next cutting. To implement iterative decomposition, the diagram in Fig. 5.2b has the following variable ordering. The nodes connected with bound functions are placed above a new cutting line. The extract above this cutting line describes function *g1*, which is performed by a bound block (2). After replacing the top part of the diagram in Fig. 5.2b with the node associated with the bound function *g1*, the diagram in Fig. 5.2c is gained. This diagram describes the function performed in a free block (3). As a result of this decomposition strategy, the technology mapping presented in Fig. 5.2d is obtained. This process is an example of classic iterative decomposition[2] [9, 13].

5.1.3 Multiple Decomposition—A Single Cut Method

Multiple decomposition using a single cutting line is performed similarly to iterative decomposition. In addition to entering the appropriate cutting line and replacing the upper part of the diagram with a fragment that contains vertices that represent the newly created bound functions, the main difference from iterative decomposition is that each time the variable order is changed in the created BDD diagram in a single decomposition step. The order is changed to ensure that the vertices that represent bound functions are located at the bottom part of the diagram (below the cut line from the next decomposition step). The concept of searching for multiple decomposition using BDD is illustrated in Fig. 5.3 [14].

Similar to the previous case, the basic function represented by the BDD diagram and shown in Fig. 5.3a is subjected to simple serial decomposition. The top extract describes the function implemented by a bound block (1). The function is replaced with a node associated with the single bound function *g0* (this function two cut nodes, i.e., the nodes connected with the variable x_2). As a result of this change in the top part of the diagram, the diagram shown in Fig. 5.3b is obtained. To search for multiple decomposition, the variable ordering should be changed. The nodes associated with bound functions may be below the cutting line (Fig. 5.3c). The result of cutting the diagram in Fig. 5.3c is expressed as follows: The top part of the diagram, which describes the function performed by the bound block (2), is replaced with a node that is connected with the newly created bound function *g1*. The obtained diagram, as shown in Fig. 5.3d, describes the function performed by the free block (3). As a result of this strategy, technology mapping that corresponds to multiple decomposition is obtained, as presented in Fig. 5.3e (see Footnote 2) [9].

The main difference among these strategies for searching for appropriate complex decompositions is the placement of the nodes connected with the bound functions. This placement often requires changing the variable ordering in BDD and substantially influences the number of cut nodes. In addition to the number of cut nodes,

---

[2]© Reprinted from Opara et al. [9], Copyright (2020), with permission from Elsevier.

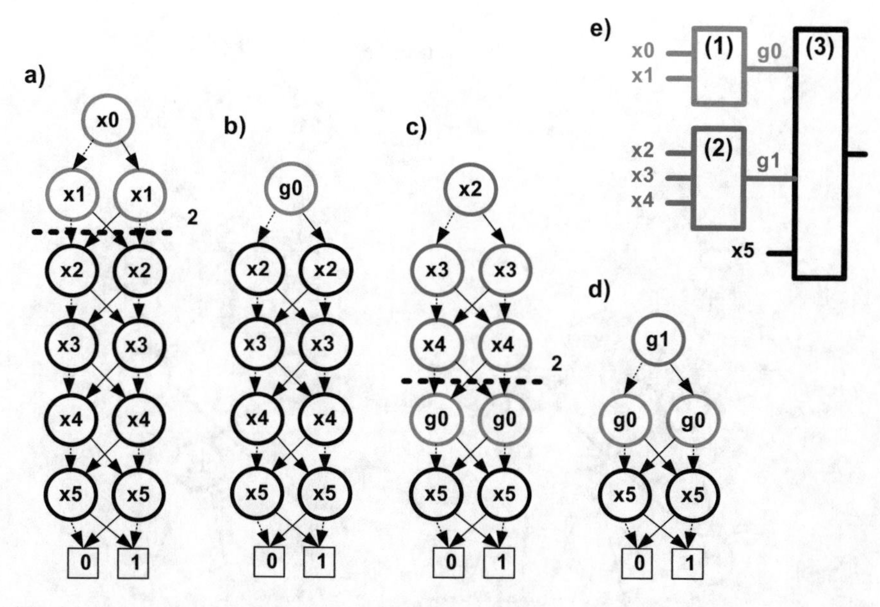

Fig. 5.3 Multiple decomposition: **a–d** show the initial steps; **e** shows the implementation of function by means of LUTs (see Footnote 2) [9]

determining the ability to map a given diagram's extract using the real logic resources of an FPGA is vital (see Footnote 2). [9].

5.2 Methods of Realizing Function Decomposition Using Multiple Cuts of the BDD Diagram

Another method for implementing cutting is to introduce additional BDD diagram cut lines. As a result of cutting the BDD diagram, BDD extracts are created and located between the cutting lines. To determine the method of determining the number of bound functions that are necessary to replace a given slice, determining the form of BDD taken by the extract between two cutting lines becomes crucial. In general, the obtained BDD extracts may have at least one root and single or multibit leaves. If they have one root and more than two multibit terminal nodes, we refer to them as MTBDDs [16–18]. However, if they have more than one root but only two terminal nodes, they are referred to as SBDDs [19–21]. When more than one root and more than two multibit terminal nodes exist in a given extract, the BDD extract can be described as a shared multiterminal BDD (SMTBDD) [22–25].

We consider a double cutting of a diagram presented in Fig. 5.4. As a result of this process, three segments—an MTBDD, SMTBDD, and SBDD—are obtained. The upper cut line designates extract 0, contains a single root (x_0) and points to 3 cut

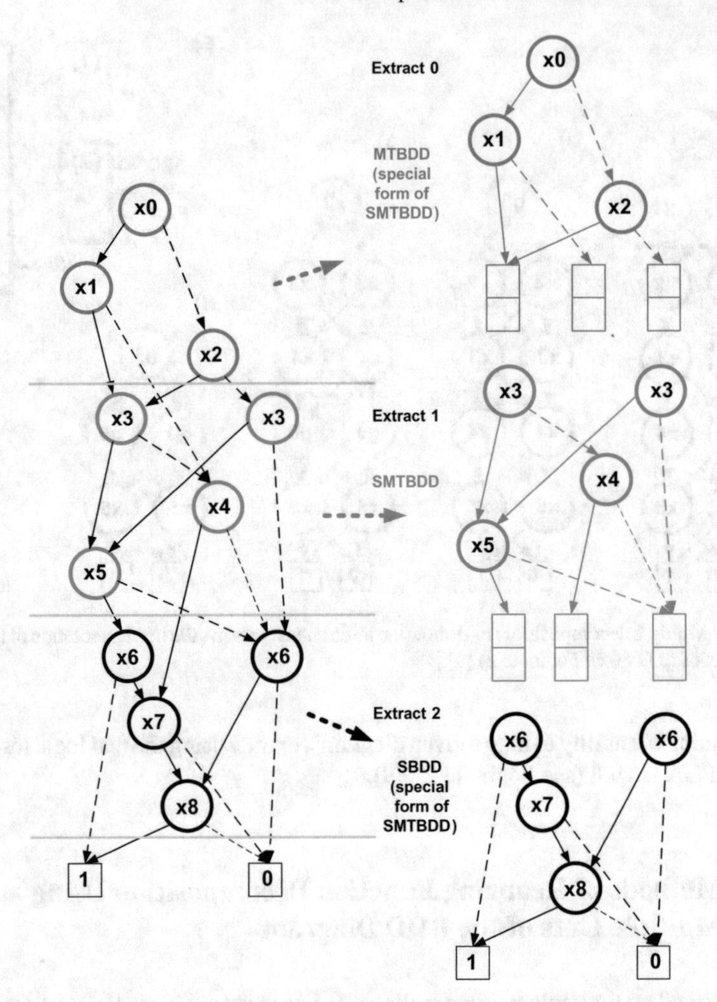

Fig. 5.4 Multiple ROBDD cutting—various forms of extracts

nodes in slice 1. Two bits are necessary to distinguish them; thus, the diagram that leaves from extract 0 will be 2 bits. Because the extract 0 diagram has a single root and several multibit leaves, this diagram is MTBDD. Two roots (x_3) and 3 two-bit leaves exist in extract 1, which means that extract 1 is SMTBDD. Extract 2 has two roots (x_6) and two 1-bit leaves; thus, extract 2 is SBDD.

MTBDD and SBDD are specific forms of SMTBDD. In general, MTBDD is SMTBDD with a single root. SBDD is SMTBDD with 2 single-bit leaves. SMTBDD can be split into several MTBDDs.

The idea of 'splitting' an SMTBDD diagram into two MTBDD diagrams is shown in Fig. 5.5.

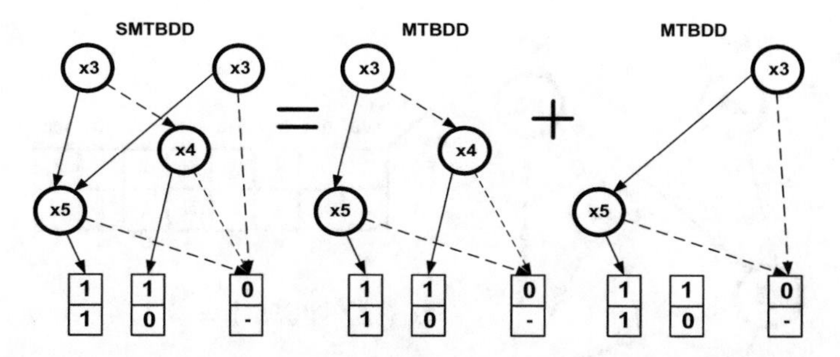

Fig. 5.5 SMTBDD as a symbolic sum of MTBDDs (see Footnote 1) [5]

5.2.1 SMTBDD in the Decomposition Process

From the point of view of performing decomposition, determining the number of bound functions needed to replace a given SMTBDD) is crucial. In the case of MTBDD, to determine the number of bound functions, determining the number of cut nodes is sufficient. For SMTBDD (and SBDD) things get complicated due to the larger number of roots. To present the method of determining the number of bound functions, in the case of a multiroot diagram, introducing a few definitions is necessary.

Definition 5.1. Shared Multi-Terminal Binary Decision Diagram (SMTBDD) is a binary decision diagram that includes n-roots and p m-bit leaves ($p \leq 2^{\mathrm{m}}$) (see Footnote 1) [5].

Definition 5.2. The root table of SMTBDD is a two-dimensional table in which columns are associated with appropriate input vectors and rows correspond to the roots of SMTBDD. The cells of the root table are associated with the symbols of cut nodes for the analyzed part (see Footnote 1) [5].

Definition 5.3. Column multiplicity of a root table, which is specified as $\nu(X_f|X_b)$, is the number of various column patterns of a root table of SMTBDD (see Footnote 1) [5].

Definition 5.4. The vector of cut nodes is a root array associated with a single root BDD diagram reduced to a single row.

Each SMTBDD may be described by means of a root table. The columns of a root table are associated with SMTBDD paths. The choice of a root and a path in SMTBDD unequivocally specifies a leaf. The choice of a column (i.e., a path in SMTBDD) and a row (i.e., a root) specifies a cell of a root table, in which a symbol related to a leaf is included. The contents of appropriate columns create column patterns. The column multiplicity of a root table is equal to the number of various column patterns (see Footnote 1) [5].

The idea of specifying column multiplicity is presented in Fig. 5.6.

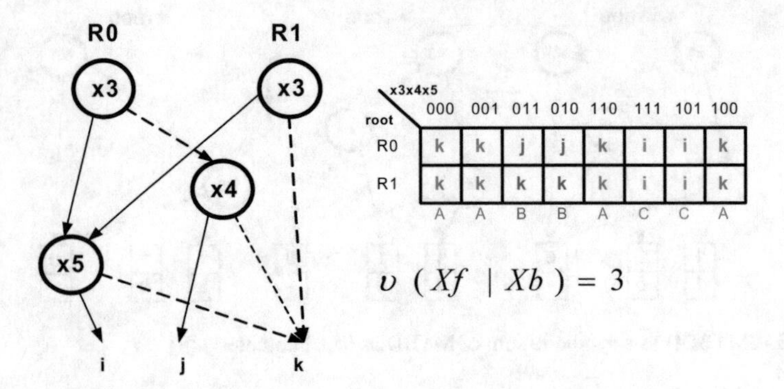

Fig. 5.6 SMTBDD diagram with a root table—specifying column multiplicity (© Reprinted from Kubica et al. [5], Copyright (2020), with permission from Elsevier)

Column multiplicity of a root table corresponds with column multiplicity of Karnaughs' map. In this case, 2 bits are needed to distinguish 3 column patterns (two bound functions).

5.2.2 *Simple Serial Decomposition—Multiple Cutting Method*

The multiple cutting method can be used to perform simple serial decomposition. In contrast to the single-cut approach (Fig. 5.7a), where the bound set was associated with the fragment above the cut line, in the case of a multiple cut line, the bound set is associated with the extract between adjacent cut lines. As shown in Fig. 5.7b, the extract between cutting lines 1 and 2 is associated with the bound set, while the extract above the cutting lines 1 and 2 are associated with the free set.

5.2.3 *Multiple Decomposition—Multiple Cutting Method*

Searching for multiple decomposition by the BDD multiple cut method is very similar to a repeated search for a simple serial decomposition. In this method, more than one extract is associated with the corresponding bound sets. As shown in Fig. 5.8, two fragments of the diagram have been associated with bound sets, i.e., extract located between the cut lines and the extract located above the upper cut line. The extract below the lower cutting line was associated with the free set. Determining the number of necessary bound functions for extracts associated with bound sets becomes necessary [14, 26].

To approximate the idea of searching for multiple decomposition by the method of multiple cutting, we consider the example in Fig. 5.9.

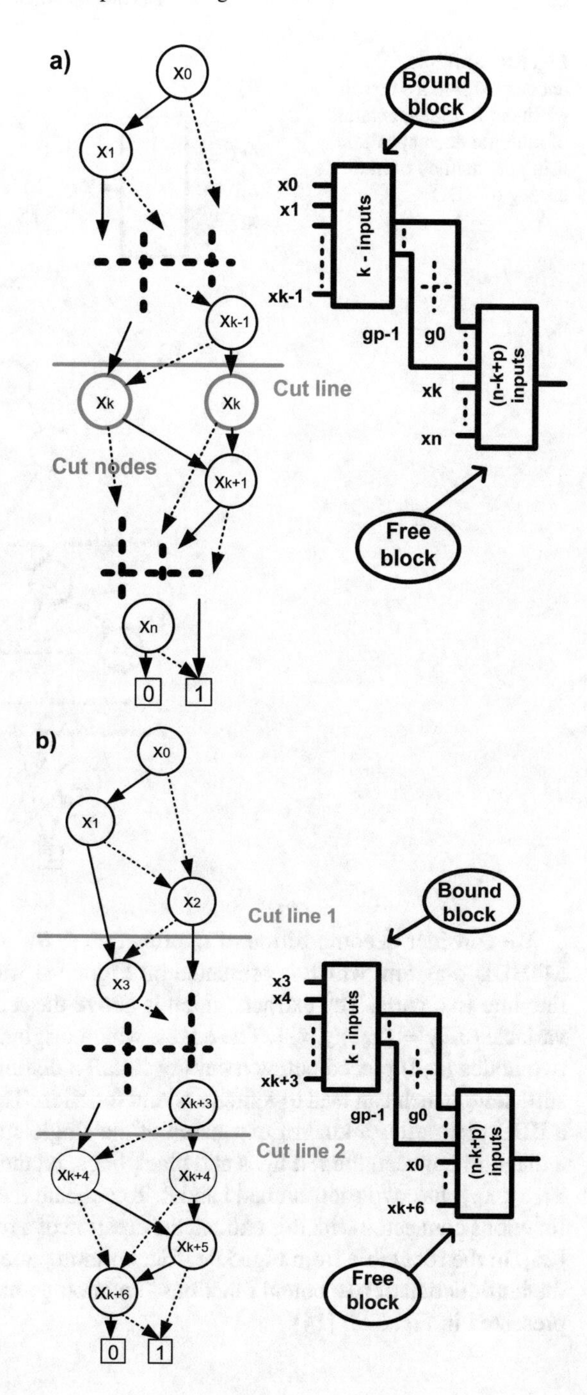

Fig. 5.7 Simple serial decomposition: **a** perform using the method of a single cutting of a BDD diagram, and **b** perform using the method of multiple cutting of a BDD diagram (© (2020) IEEE. Reprinted with permission from Opara et al. [10])

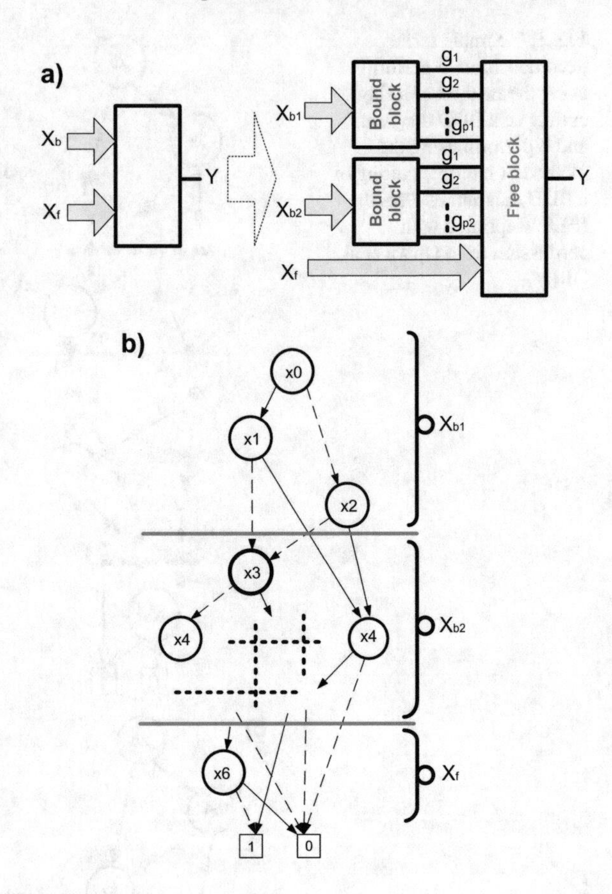

Fig. 5.8 Multiple decomposition: **a** a circuit partition; **b** implementation of multiple decomposition using the method of multiple cutting of BDD

We consider decomposition of the function f: $B^7 \rightarrow B^2$ shown in Fig. 5.9. The MTBDD diagram, which is presented on Fig. 5.9a, was divided with a red cutting line into two parts. The extract, which is above the red cutting line, includes three variables $X_{b0} = \{x_0, x_1, x_2\}$. The edges, which originate from this extract, indicate two nodes (k, l) placed below a cutting line. To distinguish them, one single bit is sufficient, which can lead to a single bound function. The multiple cutting method of a BDD diagram is a kind of expansion of the single cutting method. The extract of a diagram between the red lines and black lines, including the variables $X_{b1} = \{x_3, x_4, x_5, x_6\}$, has two roots named l and k. To indicate the number of necessary bound functions connected with this extract, the creation of a root table is needed (Fig. 5.9c) [26]. In the root table from Fig. 5.9c, four column patterns exist, which may prompt the introduction of two bound functions. Technology mapping in LUT_4/1 blocks is presented in Fig. 5.9b [14].

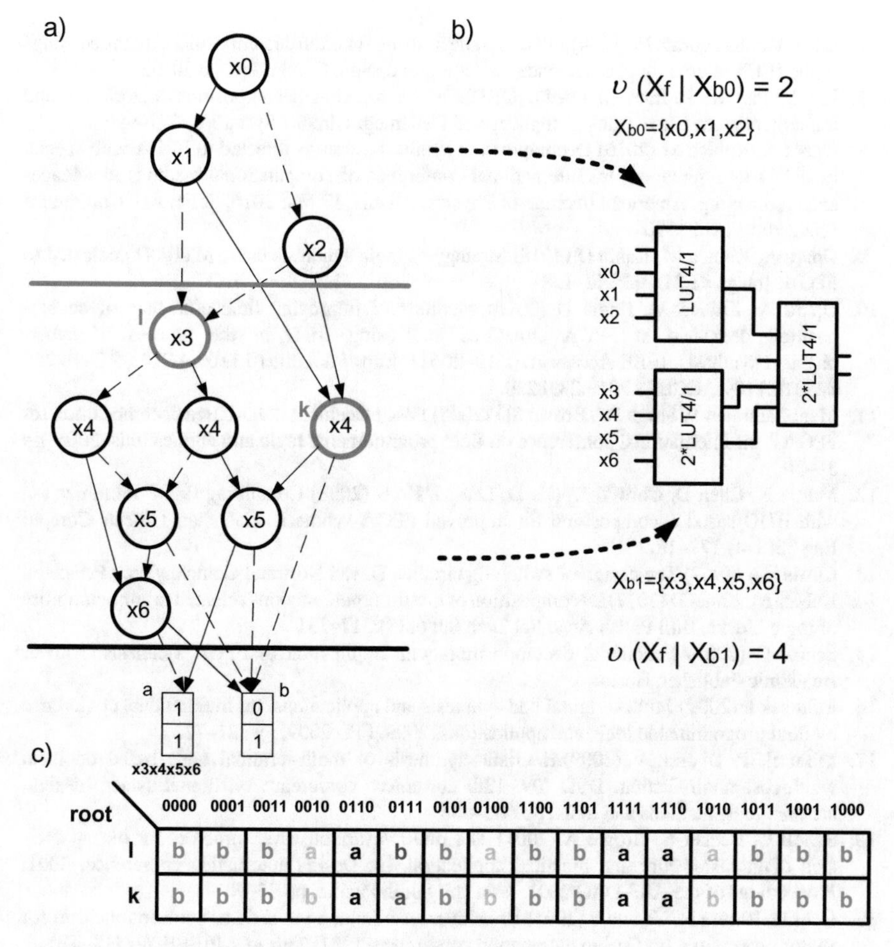

Fig. 5.9 Function decomposition described in the form of BDD, which is the result of multiple cutting

References

1. Chang S, Marek-Sadowska M, Hwang T (1996) Technology mapping for TLU FPGA's based on decomposition of binary decision diagrams. IEEE Trans Comput-Aided Des 15(10):1226–1235
2. Cheng L, Chen D, Wong M (2008) DDBDD: Delay-driven BDD synthesis for FPGAs. IEEE Trans Comput Aided Des Integr Circuits Syst 27(7):12003–21213
3. Cong J, Ding Y (1994) FlowMap: an optimal technology mapping algorithm for delay optimization in lookup-table based FPGA designs. IEEE Trans Comput Aided Des Integr Circuits Syst 1–12
4. Kubica M, Kania D (2016) SMTBDD: new form of BDD for logic synthesis. Int J Electron Telecommun 62(1):33–41
5. Kubica M, Opara A, Kania D (2017) Logic synthesis for FPGAs based on cutting of BDD. Microprocess Microsyst 52:173–187

6. Lai Y, Pan K, Pedram M (1994) FPGA synthesis using function decomposition. In: Proceedings of the IEEE international conference on computer design, Cambridge, pp 30–35
7. Lai Y, Pan K, Pedram M (1996) OBDD-based function decomposition: algorithms and implementation. IEEE Trans Comput Aided Des Integr Circuits Syst 15(8):977–990
8. Opara A, Kubica M (2016) Decomposition synthesis strategy directed to FPGA with special MTBDD representation. In: International conference of computational methods in sciences and engineering. American Institute of Physics, Athens, 17 Mar 2016, Seria: AIP Conference Proceedings, vol 1790
9. Opara A, Kubica M, Kania D (2018) Strategy of logic synthesis using MTBDD dedicated to FPGA. Integr VLSI J 62:142–158
10. Opara A, Kubica M, Kania D (2019) Methods of improving time efficiency of decomposition dedicated at FPGA structures and using BDD in the process of cyber-physical synthesis. IEEE Access 7:20619–20631. https://doi.org/10.1109/ACCESS.2019.289823010.1109/ACCESS.2019.2898230
11. Manohararajah V, Singh DP, Brown SD (2005) Post-placement BDD-based decomposition for FPGAs. In: International conference on field programmable logic and applications, 2005, pp 31–38
12. Muma K, Chen D, Choi Y, Dodds D, Lee M, Ko S (2008) Combining ESOP minimization with BDD-based decomposition for improved FPGA synthesis. Canadian J Electr Comput Eng 33(3–4):177–182
13. Curtis HA (1962) The design of switching circuits. D. van Nostrand Company Inc., Princeton
14. Kubica M, Kania D (2017) Decomposition of multi-output functions oriented to configurability of logic blocks. Bull Polish Acad Sci Tech Sci 65(3):317–331
15. Scholl C (2001) Functional decomposition with application to FPGA synthesis. Kluwer Academic Publisher, Boston
16. Mikusek P (2009) Multi-terminal bdd synthesis and applications. In: International conference on field programmable logic and applications, 2009. FPL 2009, pp 721–722
17. Mikusek P, Dvorak V (2009).Heuristic synthesis of multi-terminal bdds based on local width/cost minimization. DSD '09. 12th euromicro conference on digital system design, architectures, methods and tools, pp 605–608
18. Scholl C, Becker B, Brogle A (2001) The multiplevariable order problem for binary decision diagrams: theory and practical application. In: Design automation conference, 2001. Proceedings of the ASP-DAC 2001. Asia and South Pacific, pp 85–90
19. Ochi H, Ishiura N, Yajima S (1991) Breadth-rstmanipulation of sbdd of boolean functions for vector processing. In: Design automation conference, 1991. 28th ACM/IEEE, pp 413–416
20. Minato S, Ishiura N, Yajima S (1990) Shared binary decision diagram with attributed edges for efficient boolean function manipulation. In: Design automation conference, 1990. Proceedings, 27th ACM/IEEE, pp 52–57
21. Thornton M, Williams J, Drechsler R, Drechsler R, Wessels D (1999) Sbdd variable reordering based on probabilistic and evolutionary algorithms. In: IEEE Pacific Rim conference on communications, computers and signal processing, 1999, pp 381–387
22. Babu HMH, SASAO T (1998) Shared multi-terminal binary decision diagrams for multiple-output functions. IEICE Trans Fundamentals Electron Commun Comput Sci 81(12):2545–2553
23. Kubica M (2014) Dekompozycja i odwzorowanie technologiczne z wykorzystaniem binarnych diagramów decyzyjnych, PhD thesis, Silesian University of Technology, Gliwice, Poland
24. Kubica M, Kania D (2015) New concept of graph for function decomposition. In: PDES 2015. IFAC conference on programmable devices and embedded systems, 2015, pp 61–66
25. Kubica M, Kania D, Opara A (2016) Decomposition time effectiveness for various synthesis strategies dedicated to FPGA structures. In: International conference of computational methods in sciences and engineering, American Institute of Physics, Athens, 17 Mar 2016, Seria: AIP Conference Proceedings; vol 1790
26. Kubica M, Kania D (2017) Area-oriented technology mapping for LUT-based logic blocks. Int J Appl Math Comput Sci 27(1):207–222

Chapter 6
Ordering Variables in BDD Diagrams

Ordering variables significantly affects the number of nodes in the BDD diagram. Many known methods of ordering limit the number of BDD nodes [1–5, 16–19, 21, 22]. Unfortunately, the minimum number of nodes in the diagram does not guarantee finding the best function decomposition.

From a decomposition point of view, the selection of BDD cutting levels is key. Regardless of the method used for decomposition (single or multiple cutting), the selection of these levels depends on the degree of configurability of the logic blocks (number of inputs k) in which the function is to be mapped. This technology mapping was shown in the papers [9, 14].

The variable ordering affects the number of cut nodes and a column multiplicity of a root table, and hence the number of bound functions, as shown in Fig. 6.1. Depending on the order, variables can be assigned to different sets, which leads to different decompositions [15].

In the example in Fig. 6.1, more cut nodes occur for the order of variables, for which the graph contains more nodes. Sometimes the opposite situation can be observed, i.e., for a graph with more nodes, there is a smaller number of cut nodes. An example of the function for which this situation occurs is shown in Fig. 6.2.

It is very difficult to pre-determine the best variable ordering (NP-hard problem). It becomes necessary to analyze various variable orderings, and choose the one that will provide the best decomposition. Typically, the number of variables for functions described with BDD is large, which may lead to the need to analyze a large number of variable orderings. In this situation, the synthesis time can be extremely long. Therefore, there is a need to develop methods for searching for good variable orderings that will be time-efficient.

It is necessary to specify some initial variable ordering. It turns out that good results are obtained by adopting such a variable ordering in which the variables on which the function depends strongly are placed closest to the BDD root. Often, after completing the search for effective ordering, you get a variable ordering that is very similar to initial the variable ordering. An initial variable ordering is gained by

M. Kubica et al., *Technology Mapping for LUT-Based FPGA*, Lecture Notes in Electrical Engineering 713, https://doi.org/10.1007/978-3-030-60488-2_6

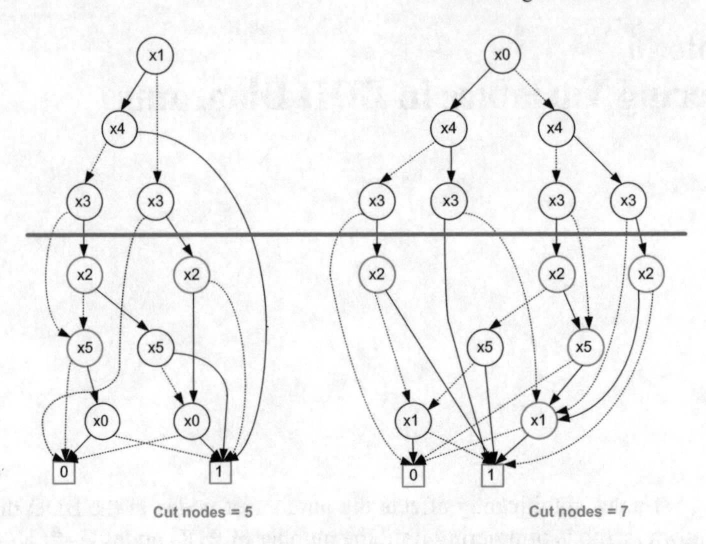

Fig. 6.1 Influence of a variable ordering in BDD on the number of cut nodes (© (2020) IEEE. Reprinted, with permission, from Opara et al. [15])

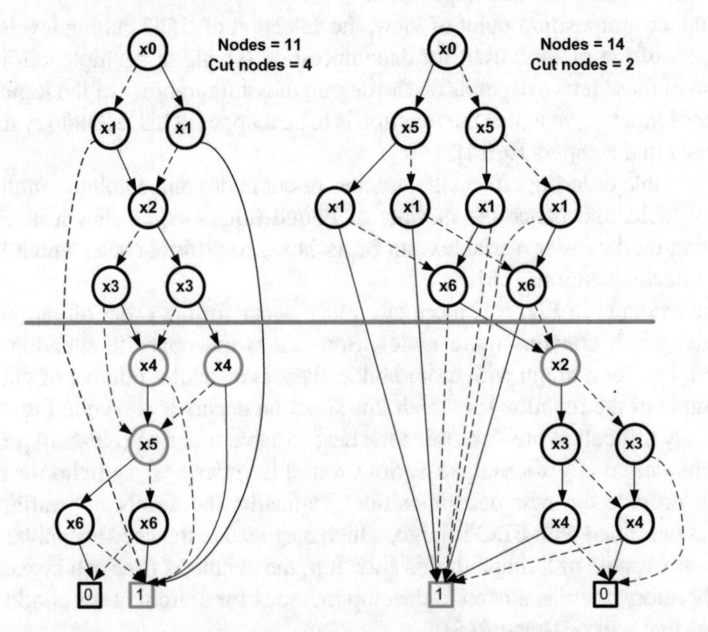

Fig. 6.2 Influence of a variable ordering in BDD on the number of cut nodes (more nodes but less cut nodes) (© (2020) IEEE. Reprinted, with permission, from Opara et al. [15])

analyzing the frequency of appearance of separate variables in the expression, which is in the form of the sum of products [15].

The idea of variable ordering search is slightly different for single and multiple cut methods. In both cases, the levels of cutting a diagram should be determined first, which leads to the determination of the number of separate bound sets *card*(*Xb*) and a free set *card*(*Xf*) [15].

Each node in the diagram is assigned a unique ID that specifies the level on which a given node is located [6]. The structure data that is referred to as the bdd_manager [13, 20] contains information common to several diagrams such as: operation cache, index tables. In addition, for a single bdd_manager, all functions must have the same variable ordering.

During decomposition, some diagrams must be stored with different variable orderings. It became necessary to develop appropriate methods that would allow easy displacement of nodes below and above the cutting line [15].

Let's first consider the case using a single cutting method. In the BDD creation process, we assume that the number of variables is twice as high as the number of variables of a function. In the classical approach, each variable is assigned a single ID number. The introduced modification consists in the fact that a single variable is assigned two ID numbers, one of which is even (position below the cutting line) and the other odd (position above the cutting line). Depending on the set to which an analyzed variable is ascribed, only one ID that describes this variable is 'active' in a given moment. Moving the variable between the upper and lower parts of BDD therefore comes down to changing the assigned ID, for which you can use the function bdd_swap [12]. From the point of view of decomposition, it does not matter where in the upper (or lower) part of the diagram the considered variable is located, the key is whether it is above (or below) the cutting line [15].

In the next decomposition steps with BDD, nodes associated with the bound functions g appear. As before, two IDs are also associated with g variables. And the idea of moving them between the lower and upper parts of the diagram is exactly the same as before. The idea of transferring variables between sets using BDD is illustrated in Fig. 6.3 [15].

Let's then consider the case using the multiple cut method. In this case, for the function of n variables, it is necessary to create such a diagram that will take into account the number of n^2 variables. This approach leads to the need to check a much larger number of orderings than was the case with a single cutting line. The essence of this method is that the number of bound sets is greater than 1, which leads to multiple decomposition. It is important to determine, not only whether a given variable is included in the bound set, it should be specified in which of the bound sets the variable will be included. As shown in [15], each of the n variable functions can be associated with n ID numbers, which leads to a symbolic division of the diagram into levels in which there are nodes associated with only one variable. Each level includes the set of n ID numbers. One of these numbers is active in a given moment. Depending on what position in the BDD the given variable is to be placed, one of the ID numbers associated with this variable is activated in the ID set at a given diagram level. Each variable includes an ID number on every level of BDD.

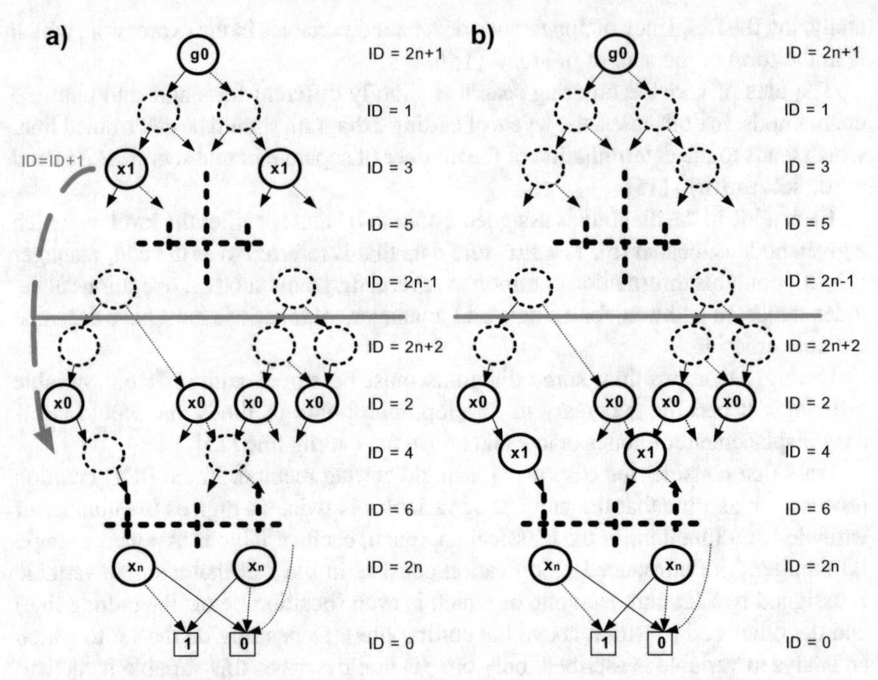

Fig. 6.3 Relocating the variable x_1 from a bound set to a free set: a) a symbolic BDD diagram before relocating, and b) a symbolic BDD diagram after relocating (© (2020) IEEE. Reprinted, with permission, from Opara et al. [15])

As before, moving a variable in BDD comes down to changing the ID number. This method is presented in Fig. 6.4 [15].

Consider the case in which the variable ordering occurs such that x_1 occurs above x_2 as shown in Fig. 6.4a. In this situation, variable x_1 is assigned ID = 1 (marked with a continuous line on the left side of BDD). The variable x_2 is assigned ID = n + 2. Changing the order of the variables, as shown in Fig. 6.4b, the variable x_2 is above the variable x_1, which is associated with the change in the respective ID values. For variable x_2 the value ID = 2, while for x_1 the value ID = n + 1, which leads to the situation that the variable data is active at other levels in the diagram [15].

The method of ID modification can be determined using Eq. (6.1):

$$\mathrm{ID} = (level - 1) * numb\_of\_var + idx, \tag{6.1}$$

where level—a level on which an analyzed variable has to be placed in BDD, numb_of_var—the number of variables placed in BDD, and idx—the index of a variable to relocate (numbering from 1).

The IDs of variables g (bound functions) [15] are assigned in the same way.

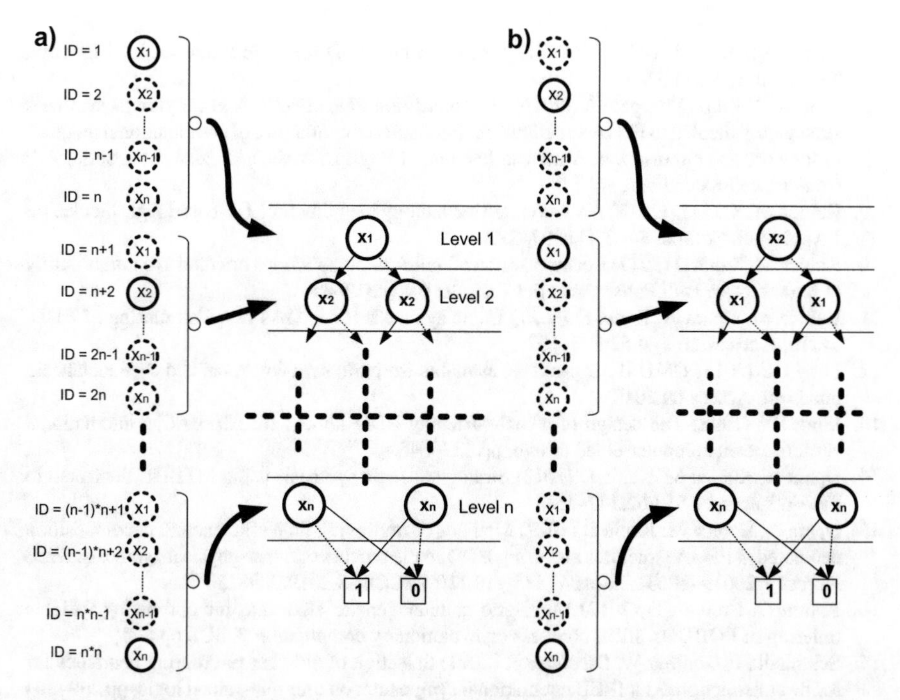

Fig. 6.4 Change of a variable ordering: **a** an ordering in which x_1 is associated with a BDD root, and **b** an ordering in which x_2 is associated with a BDD root (© (2020) IEEE. Reprinted, with permission, from Opara et al. [15])

Both methods of changing variable orderings were implemented in academic synthesis tools. In the case of a single cutting method, dekBDD is employed [8, 11, 14, 15]. In the case of the multiple cutting method, MultiDec is utilized [7, 9, 10].

References

1. Chaudhury S, Dutta A (2011) Algorithmic optimization of BDDs and performance evaluation for multi-level logic circuits with area and power trade-offs. Circuits Syst 02:217–224
2. Ebendt R (2003) Reducing the number of variable movements in exact BDD minimization. IEEE Int Symp Circuits Syst (ISCAS) 5:V605–V608
3. Ebendt R, Gunther W, Drechsler R (2004) Combining ordered best-first search with branch and bound for exact BDD minimization. In: Design automation conference 2004. Proceedings of the ASP-DAC 2004. Asia and South Pacific, pp 876–879
4. Hong Y, Beerel PA, Burch JR, McMillan KL (2000) Sibling-substitution-based BDD minimization using don't cares. Trans Comput-Aided Des Integr Circuits Syst 19(1):44–55
5. Gomez-Prado D, Ren Q, Askar S, Ciesielski M, Boutillon E (2004) Variable ordering for Taylor expansion diagrams, High-level design validation and test workshop 2004. Ninth IEEE international high-level design validation and test workshop, pp 55–59
6. Janssen G, A consumer report on BDD packages, 16th symposium on integrated circuits and systems design (SBCCI'03)

7. Kubica M, Kania D (2016) SMTBDD: new form of BDD for logic synthesis. Int J Electron Telecommun 62(1):33–41
8. Kubica M, Kania D, Opara A (2016) Decomposition time effectiveness for various synthesis strategies dedicated to FPGA structures. In: International conference of computational methods in sciences and engineering. American Institute of Physics, Athens, 17 Mar 2016, Seria: AIP Conference Proceedings, vol 1790
9. Kubica M, Kania D (2017) Area-oriented technology mapping for LUT-based logic blocks. Int J Appl Math Comput Sci 27(1):207–222
10. Kubica M, Kania D (2017) Decomposition of multi-output functions oriented to configurability of logic blocks. Bull Polish Acad Sci Tech Sci 65(3):317–331
11. Kubica M, Opara A, Kania D (2017) Logic synthesis for FPGAs based on cutting of BDD. Microproces Microsyst 52:173–187
12. Long DE (2018) CMU BDD package, available from https://www.cs.cmu.edu/~modelcheck/bdd.html. Access 09 2018
13. Long DE (1998) The design of a cache-friendly BDD library. In: IEEE/ACM international conference on computer-aided design, pp 639–645
14. Opara A, Kubica M, Kania D (2018) Strategy of logic synthesis using MTBDD dedicated to FPGA. Integr VLSI J 62:142–158
15. Opara A, Kubica M, Kania D (2019) Methods of improving time efficiency of decomposition dedicated at FPGA structures and using BDD in the process of cyber-physical synthesis. IEEE Access 7:20619–20631. https://doi.org/10.1109/ACCESS.2019.2898230
16. Rotaru C, Brudaru O (2012) Multi-grid cellular genetic algorithm for optimizing variable ordering of ROBDDs. IEEE congress on evolutionary computation (CEC), pp 1–8
17. Schmiedle F, Gunther W, Drechsler R (2001) Selection of efficient re-ordering heuristics for MDD construction. 31st IEEE international symposium on multiple-valued logic, pp 299–304
18. Sharma P, Singh N (2014) Improved BDD compression by combination of variable ordering techniques. International conference on communication and signal processing, pp 617–621
19. Siddiqui M, Ahmad SN, Beg MT (2016) Modified GA method for variable ordering in BDD for MIMO digital circuits. In: IEEE international conference on advances in electronics communication and computer technology (ICAECCT), pp 378–382
20. Somenzi F (2018) Colorado University Decision Diagram package (CUDD), cudd-3.0.0.tar.gz available from https://github.com/ivmai/cudd. Access 09 2018
21. Thornton MA, Williams JP, Drechsler R, Drechsler N (1999) Variable reordering for shared binary decision diagrams using output probabilities. Design automation and test in Europe conference and exhibition, pp 758–759
22. Varma Ch (2018) An enhanced algorithm for variable reordering in binary decision diagrams. In: 9th international conference on computing communication and networking technologies (ICCCNT), pp 1–4

Chapter 7
Nondisjoint Decomposition

In the process of decomposition, various models of disjoint decomposition, for which $Xb \cap Xf = \Phi$, are applied. The decomposition process can be optimized. Nondisjoint decomposition produces better results where the area is concerned [1–6]. Nondisjoint decomposition reduces the area in bound blocks [7, 8].

In general, nondisjoint decomposition is a kind of expansion of serial decomposition (disjoint). One part of the variables may be included in both a bound set and a free set. Thus, the third set of variables Xs, including common variables (7.1), may be distinguished [9].

$$X_b \cap X_f \neq \Phi; \quad X_b \cap X_f = X_s. \tag{7.1}$$

Nondisjoint decomposition is presented in Fig. 7.1.

The process of attaching some variables to both a bound set and a free set may reduce the number of bound functions g [10] (which means you can limit the number of logic blocks needed to carry out the bound functions). The variables from a common set Xs will fulfill the role of bound functions, which means that part of the bound functions is replaced by input variables. Some variables can fulfill the role of a 'switching' function. Therefore, searching for nondisjoint decomposition is based on searching for the variables that may fulfill the role of bound functions. Nondisjoint decomposition may significantly improve the efficiency of logic resources [11–13]. This issue is the main topic of many papers. The ways of searching for nondisjoint decomposition were presented in the papers by [14, 15]. The starting point of the proposed searching nondisjoint decomposition is disjoint decomposition (e.g., a simple serial decomposition). The process of searching for an appropriate decomposition includes attaching the subsequent variables to the set Xs and checking whether the attachment of a variable xi to the set Xs will reduce the number of bound functions g. A classic method that checks this condition is based on the analysis of the vectors of cut nodes obtained for a stable value of a 'switching' variable (variable x_0), as shown in the example [9].

M. Kubica et al., *Technology Mapping for LUT-Based FPGA*, Lecture Notes in Electrical Engineering 713, https://doi.org/10.1007/978-3-030-60488-2_7

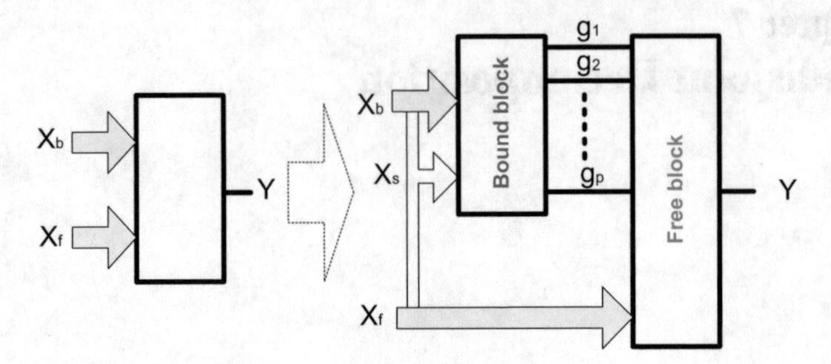

Fig. 7.1 Nondisjoint decomposition

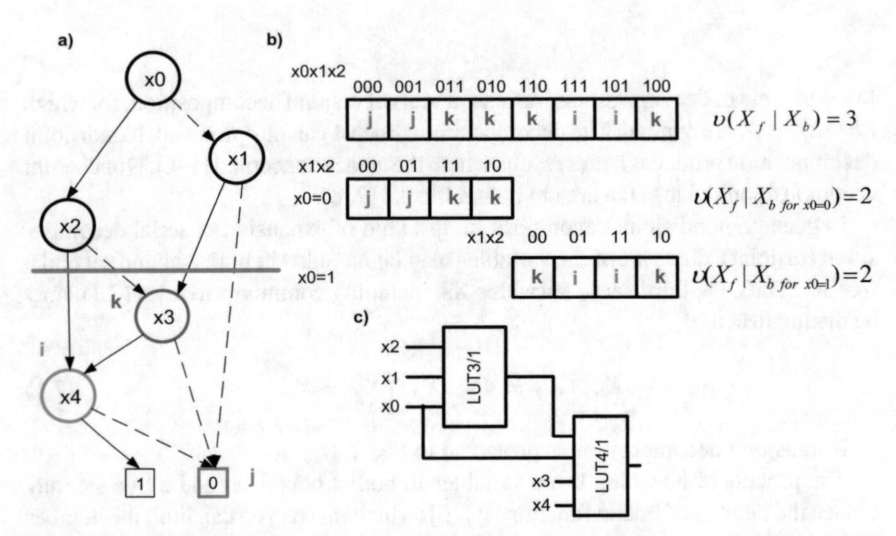

Fig. 7.2 Nondisjoint decomposition: **a** the diagram that represents the function, **b** the vectors of cut nodes, and **c** the result of technology mapping

We consider the function $f\colon B^4 \rightarrow B$ described using the ROBDD diagram presented in Fig. 7.2. We consider the cutting of a diagram on the third level counting from a root (Fig. 7.2a). The three cut nodes i, j, and k exist for the analyzed cutting. Two bits (two bound functions) are needed to differentiate them. The vector of the cut nodes was presented in Fig. 7.2b. To minimize the number of bound functions, searching for a variable that belongs to a bound set $\{x_0, x_1, x_2\}$ that may replace one of the bound functions, should be started [9].

We consider whether the variable x_0 may fulfill the role of a bound function. Figure 7.2b presents the partition of a vector of cut nodes for the variable $x_0 = 0$ and $x_0 = 1$. As the result of partition, two four-element vectors were created. The first vector for $x_0 = 0$ includes the symbols associated with the two cut nodes j and k; thus, one bound function is necessary. This case is similar to the case of the vector connected with $x_0 = 1$, where the two symbols i and k (a single bound function) exist because the number of required bound functions for $x_0 = 0$ and $x_0 = 1$ is lower by 1 than the number of bound functions for nondisjoint decomposition. It means that the variable x_0 may fulfill the role of bound function. The variable x_0 shall be attached to both a bound block and a free block. The structure of a circuit after decomposition is shown in Fig. 7.2c [9].

In the case of multiroot SMTBDD diagrams, the method of searching for nondisjoint decomposition may be similarly defined [9, 16, 17]. Disjoint decomposition is the starting point of searching for nondisjoint decomposition associated with a given extract. All variables that belong to SMTBDD diagram are analyzed where the possibility of replacing bound functions is concerned. This process is based on the attachment of variables to the set *Xs* and involves checking whether it is profitable. If the attachment of the variable *xi* to the set *Xs* reduces the number of bound functions g, the variable xi will remain in the set *Xs*.

This case is similar to the case of a one-root diagram, where the development of a method that will decide whether the attachment of the variable *xi* to the set *Xs* reduces the number of bound functions is essential.

Each variable xi corresponds to the nodes in the SMTBDD diagram that are placed on a given level. The variable xi may assume the value 0 (xi = 0) or the value 1 (xi = 1), which is connected with appropriate edges from a given node. These edges indicate appropriate subdiagrams, such as the subdiagrams for xi = 0 and xi = 1 that may be distinguished. Both subdiagrams indicate a given number of cut nodes for the given roots. Root tables for xi = 0 and xi = 1, for which the column multiplicity is determined, can be created. The number of various column patterns determines the number of necessary bits (bound functions) to distinguish them for both variable values xi = 0 and xi = 1. If the number of bits (bound functions) that are needed to differentiate the column patterns in a root table for the nodes indicated by the subdiagram connected with xi = 0 is smaller than the number of bits (bound functions) for nondisjoint decomposition and the number of bits (bound functions) for the subdiagram connected with xi = 1 fulfills the same condition, the variable xi may fulfill the role of a bound function. The analyzed method of searching for nondisjoint decomposition for a multiroot SMTBDD diagram is presented in the example [9].

For the function described with the diagram presented in Fig. 7.3a, the extract between two cutting lines that include three variables $\{x_2, x_3, x_4\}$, is separated. As the result of cutting, the SMTBDD diagram that has two roots a and b is created. The SMTBDD diagram is connected with four cut nodes, such as m, n, o and p (Fig. 7.3b). To determine the number of bound functions, a root table is created (Fig. 7.3c). Because the column multiplicity of a root table is 4, the creation of two bound functions is necessary [9].

To replace one of the bound functions with the variable x, searching for nondisjoint decomposition should be started. First, using the variable x_2 as a 'switching' variable is considered. Figure 7.3c shows two root tables associated with $x_2 = 0$ and $x_2 = 1$. In both cases, the column multiplicity is 2. Thus, one single bit will be sufficient for differentiating them. The number of bound functions is smaller than the number of bound functions for nondisjoint decomposition for both $x_2 = 0$ and $x_2 = 1$, which means that the variable x_2 may fulfill the role of a bound function. In the analyzed

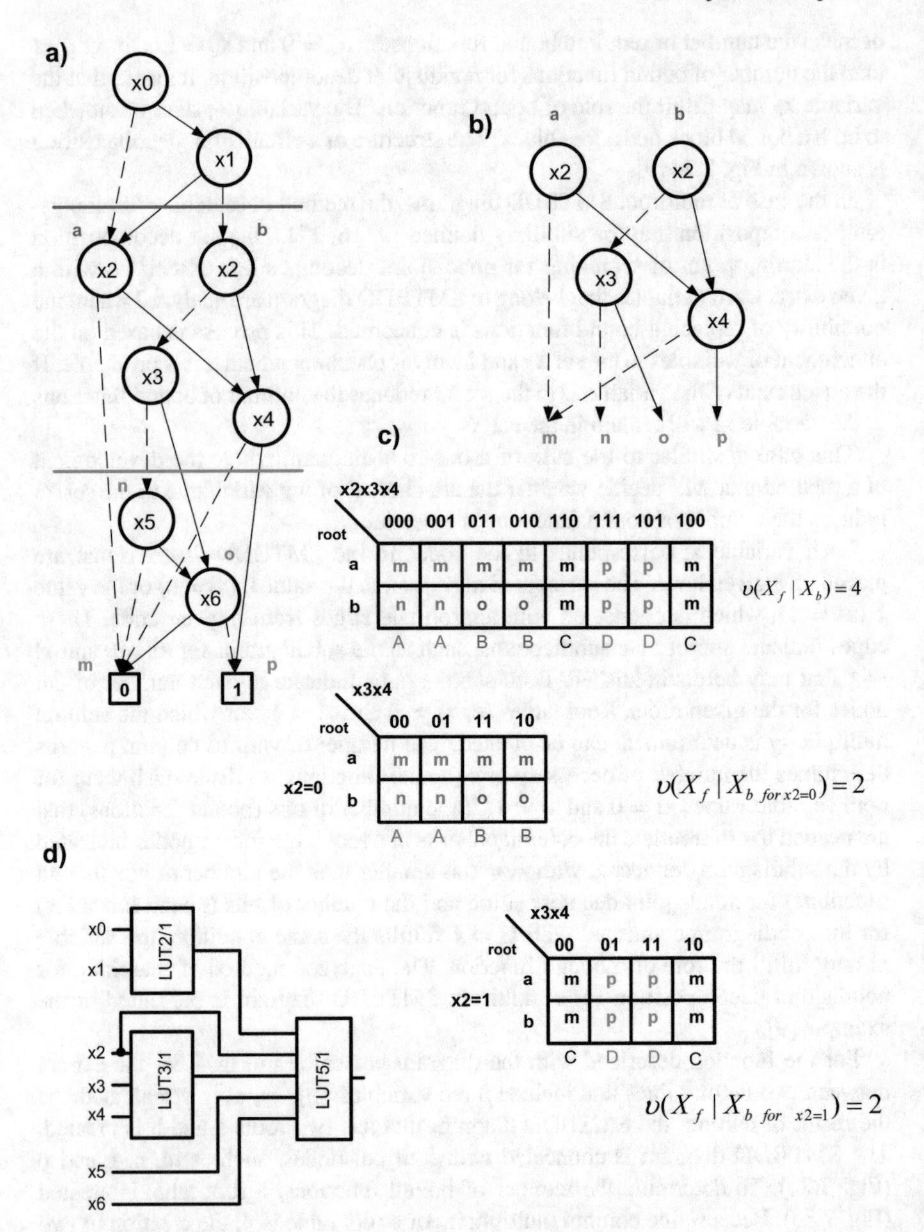

Fig. 7.3 Nondisjoint decomposition in SMTBDD diagrams; **a** ROBDD diagram with cutting lines, **b** SMTBDD diagram, **c** root tables, and **d** obtained structure

case, only one single variable x that may fulfill the role of the function g can exist. Therefore, searching for nondisjoint decomposition should be finished. The obtained structure of a circle is presented in Fig. 7.3d [9].

References

1. Hrynkiewicz E, Kołodziński S (2010) Non-disjoint decomposition of logic functions in Reed-Muller spectral domain. In: 13th IEEE symposium on design and diagnostics of electronic circuits and systems, 14–16 April 2010, pp 293–296
2. Kania D (2012) Układy Logiki Programowalnej - Podstawy syntezy i sposoby odwzorowania technologicznego. PWN, Warszawa
3. Opara A, Kubica M, Kania D (2019) Methods of improving time efficiency of decomposition dedicated at FPGA structures and using BDD in the process of cyber-physical synthesis. IEEE Access 7:20619–20631
4. Mazurkiewicz T, Łuba T (2019) Non-disjoint decomposition using r-admissibility and graph coloring and its application in index generation functions minimization. In: 26th international conference mixed design of integrated circuits and systems, MIXDES, pp 252–256
5. Rawski M, Jozwiak L, Nowicka M, Luba T (1997) Non-disjoint decomposition of Boolean functions and its application in FPGA-oriented technology mapping. In: Proceedings of the 23rd EUROMICRO conference: new frontiers of information technology, pp 24–30
6. Rawski M, Selvaraj H, Luba T (2003) An application of functional decomposition in ROM-based FSM implementation in FPGA devices. In: Proceedings, Euromicro symposium on digital system design 2003, pp 104–110
7. Opara A, Kubica M (2017) Optimization of synthesis process directed at FPGA circuits with the usage of non-disjoint decomposition. In: Proceedings of the international conference of computational methods in sciences and engineering 2017. American Institute of Physics, Thessaloniki, 21 April 2017, Seria: AIP conference proceedings; vol 1906, Art. no. 120004
8. Opara A, Kubica M (2018) The choice of decomposition taking non-disjoint decomposition into account. In: Proceedings of the international conference of computational methods in sciences and engineering 2018. American Institute of Physics, Thessaloniki, 14 Mar 2018, Seria: AIP Conference Proceedings; vol 2040, Art. no. 080010
9. Kubica M, Kania D (2017a) Decomposition of multi-output functions oriented to configurability of logic blocks. Bull Polish Acad Sci Tech Sci 65(3):317–331
10. Scholl Ch (2001) Functional decomposition with application to FPGA synthesis. Kluwer Academic Publisher, Boston
11. Dubrova E (2004) A polinominal time algorithm for non-disjoint decomposition of multi-valued functions. In: 34th international symposium on multiple-valued logic, pp 309–314
12. Dubrova E, Teslenko M, Martinelli A (2004) On relation between non-disjoint decomposition and multiple-vertex dominators, circuits and systems, 2004. ISCAS '04, vol 4, pp 493–496
13. Hrynkiewicz E, Kołodziński S (2010) An Ashenhurst Deisjoint and Non-disjoint decomposition of logic functions in reed—muller spectral domain, 17th international conference "mixed design of integrated circuits and systems", 2010, pp 293–s296
14. Opara A (2008) Decomposition synthesis methods of combinational circuits using BDD—PhD thesis, Gliwice. https://ssuise-keit.multimedia.edu.pl/doktoraty.php (in Polish)
15. Yamashita S, Sawada H, Nagoya A (1998) New methods to find optimal non-disjoint bi-decompositions. Design automation conference 1998. Proceedings of the ASP-DAC '98, pp 59–68
16. Kubica M, Kania D (2016) SMTBDD: new form of BDD for logic synthesis. Int J Electron Telecommun 62(1):33–41
17. Kubica M, Kania D (2017b) Area-oriented technology mapping for LUT-based logic blocks. Int J Appl Math Comput Sci 27(1):207–222

Chapter 8
Decomposition of Multioutput Functions Described Using BDD

In practice, multioutput functions instead of single functions are implemented. By using common dependencies for several functions, a realization that consumes a smaller number of logical resources of a programmable device than that if each function is separately treated can be achieved [1–6].

The advantages of using decomposition of a multioutput function in the process of logic synthesis include the opportunity to share logic blocks [7]. The idea of sharing a bound block is shown in Fig. 8.1.

All G functions of the bound block—full sharing (Fig. 8.1) or some partial sharing [8]—can be shared, and nondisjoint decomposition [9, 10] can be treated as a form of sharing resources. This chapter will consider building fully shared bound blocks, and Chap. 9 will consider partially shared blocks. Section 8.1 will show how to create common bound blocks using MTBDD and BDD diagrams, and Sect. 8.2 will show how to group functions into multioutput functions and set the order of variables in BDD diagrams. In the last section, we will present the effective merging and splitting of the BDD method.

8.1 Creating Common Bound Blocks

By using the common relations of several single-output functions, better technology mappings may be obtained. The synthesis of a multioutput function is usually more favorable than the synthesis of each separate single-output function regarding the usage of the area [3].

Ashenhurst–Curtis decomposition may be expanded to multioutput functions. SBDD and MTBDD can be used to represent a multioutput function [11]. The problem is the choice of the single-output functions that have to create a multi–output function and for which a common bound block will emerge (Fig. 8.2)[1] [2].

---

[1] © Reprinted from Kubica et al. [2], Copyright (2020), with permission from Elsevier.

M. Kubica et al., *Technology Mapping for LUT-Based FPGA*, Lecture Notes in Electrical Engineering 713, https://doi.org/10.1007/978-3-030-60488-2_8

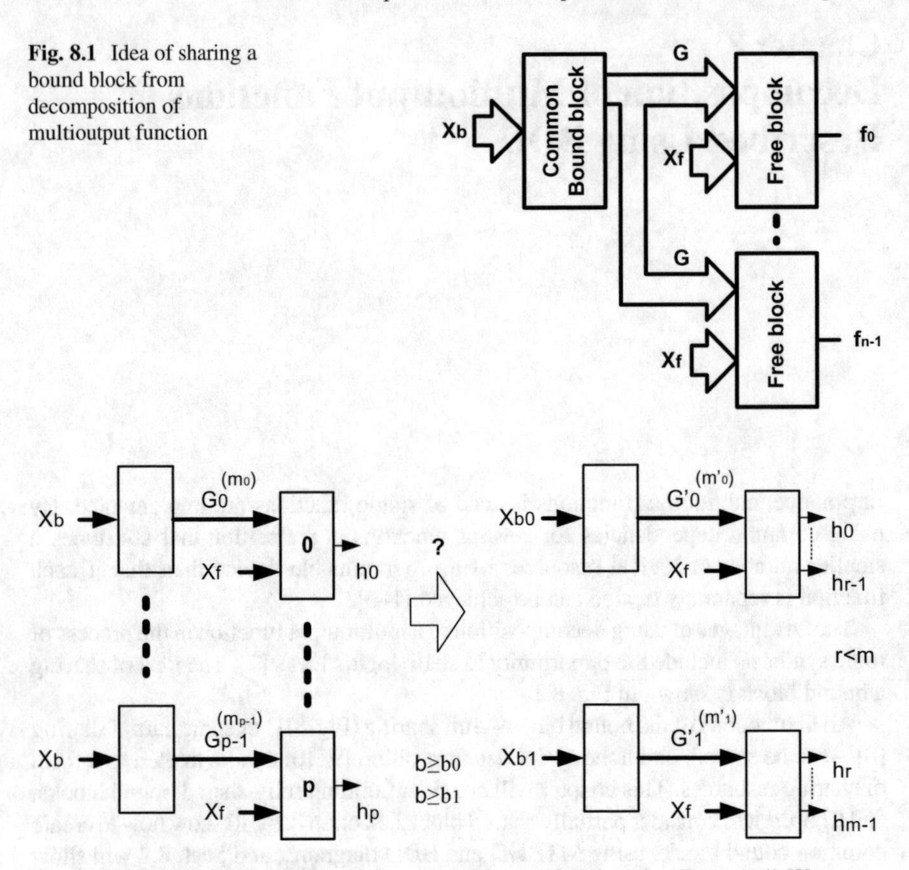

Fig. 8.1 Idea of sharing a bound block from decomposition of multioutput function

Fig. 8.2 Using common bound blocks for two multioutput functions (see Footnote 1) [2]

The ability to use common relations for several functions may cause decomposition using common bound blocks. Figure 8.3a–c illustrate the diagrams of three functions. A bound set $X_b = \{x_0, x_1, x_2\}$ was chosen. To present the relation between cut nodes in a BDD diagram and a MTBDD diagram and the methods for determining the number of column patterns, a table of cut nodes was introduced (see Footnote 1) [2].

Definition 8.1 The table of cut nodes in BDD diagrams of a multioutput function is a two-dimensional table, whose columns are related to appropriate combinations of variables that belong to the bound set and the rows that correspond to the roots of BDD diagrams of separate functions. The cells of the table are associated with cut nodes in the diagrams (see Footnote 1) [2].

Figure 8.4 shows the table of cut nodes for the separate paths x_0, x_1, and x_2. For the f function, three cut nodes exist: 0, a, and 1 (Fig. 8.3a). Therefore, two functions g_i are needed. For the f_1 function, two cut nodes exist (Fig. 8.3b); thus, one g_i function is needed similar to the case of the f_2 function. To implement all bound blocks, in the

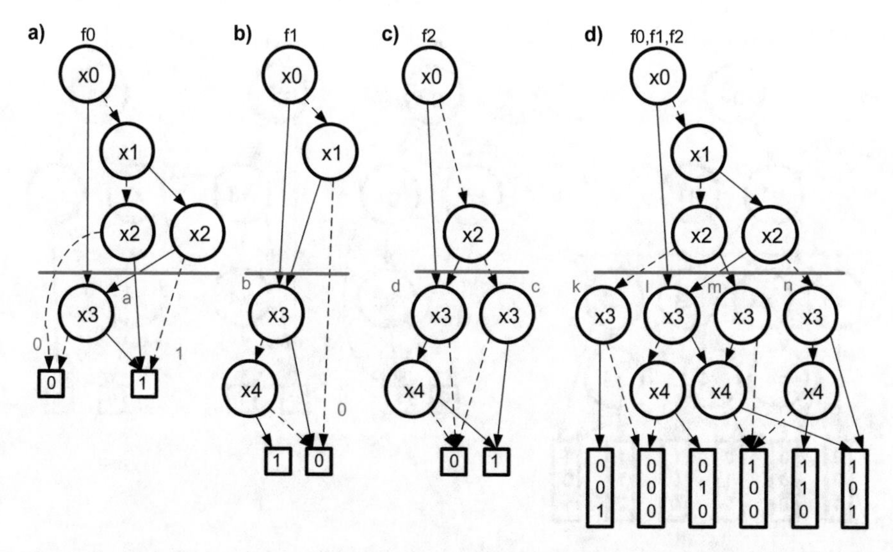

Fig. 8.3 Diagrams of functions joined into one MTBDD diagram (see Footnote 1) [2]

Fig. 8.4 Table of cut nodes for the diagrams from Fig. 8.3 (see Footnote 1) [2]

functions (roots) \ $x_0x_1x_2$	000	001	011	010	110	111	101	100
f_0	0	1	a	1	a	a	a	a
f_1	0	0	b	b	b	b	b	b
f_2	c	d	d	c	d	d	d	d
f_0,f_1,f_2	k	m	l	n	l	l	l	l

Target nodes for the path $x_0x_1x_2$

case of separately searching for decomposition for each function, four g_i functions are needed (see Footnote 1) [2].

To create a common bound block, an MTBDD diagram may be built (Fig. 8.3d). For given functions, which are treated as multioutput functions, four cut nodes exist: **k, l, m, n**. Each of these nodes corresponds to one column pattern from the table in Fig. 8.4. For these nodes, the following codes were assigned: **k** $\hat{=}00_{g0g1}$, **l** $\hat{=}$ 11_{g0g1}, **m** $\hat{=}$ 01_{g0g1}, and **n** $\hat{=}$ 10_{g0g1}. After replacing the top part of the diagram of the multioutput function—f_0, f_1, and f_2—with the part of the diagram that includes the variables g_i, the diagram of the multioutput function h_0, h_1, and h_2 is created (Fig. 8.5) (see Footnote 1) [2].

A bound block is described using two functions g_i (previously four). Decomposition of the multioutput function enabled the reduction of two of the LUTs. In some cases, the reduction in the number of blocks may be even greater. Figure 8.6 presents the function diagrams and their implementation using common bound blocks. Three functions: f_0, f_1 and f_2 are shown in Fig. 8.6a. The function was assumed to be

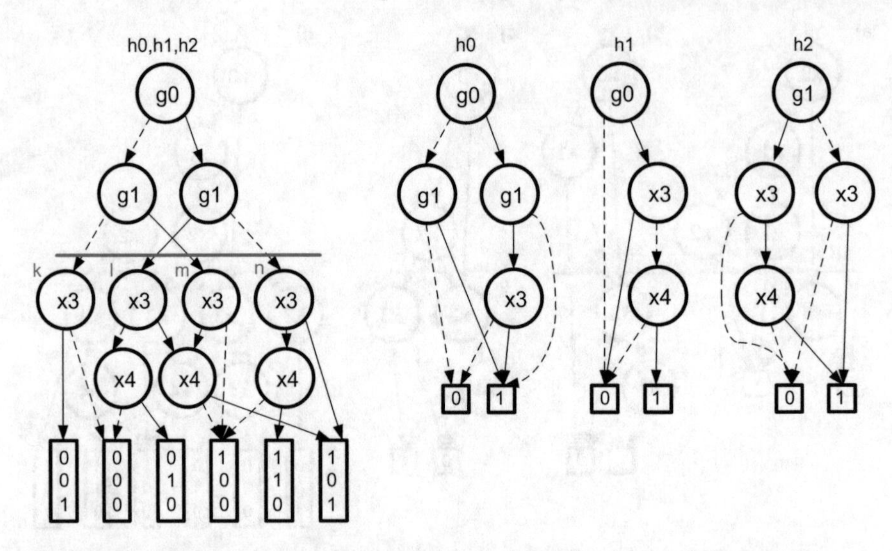

Fig. 8.5 Diagrams of multioutput function and single-output function (see Footnote 1) [2]

implemented with 3-inputs LUT blocks (LUT3/1), therefore the cutting line is at the third level from the root. After decomposition, it is necessary to use two bound functions for each of these three functions, which in turn leads to the need to use a total of 9 LUT3/1 blocks. The three diagrams of Fig. 8.6a can be replaced by a single MTBDD diagram, which as shown in Fig. 8.6b has three cut nodes. To code the cut nodes, two functions g (two LUT3/1 blocks) are sufficient. This time, however, the same two functions g may be utilized for each of the functions f_0, f_1, and f_2. In this case, the number of necessary blocks is 5, as shown in the structure in Fig. 8.6b (see Footnote 1) [2, 7].

The same process of searching for shared blocks may be conducted for multiroot SMTBDDs [12–16]. The way of proceeding is nearly the same, with the exception that the symbols, which are placed in the cells of a partition table, are connected with an appropriate column pattern in a root table and are not connected with the cut nodes.

If we have a large number of functions, merging all functions into multioutput function, which is performed using common resources, may cause a substantial increase in the number of LUTs [17]. The next section presents an algorithm of gradual merging of BDDs that represent separate functions [18].

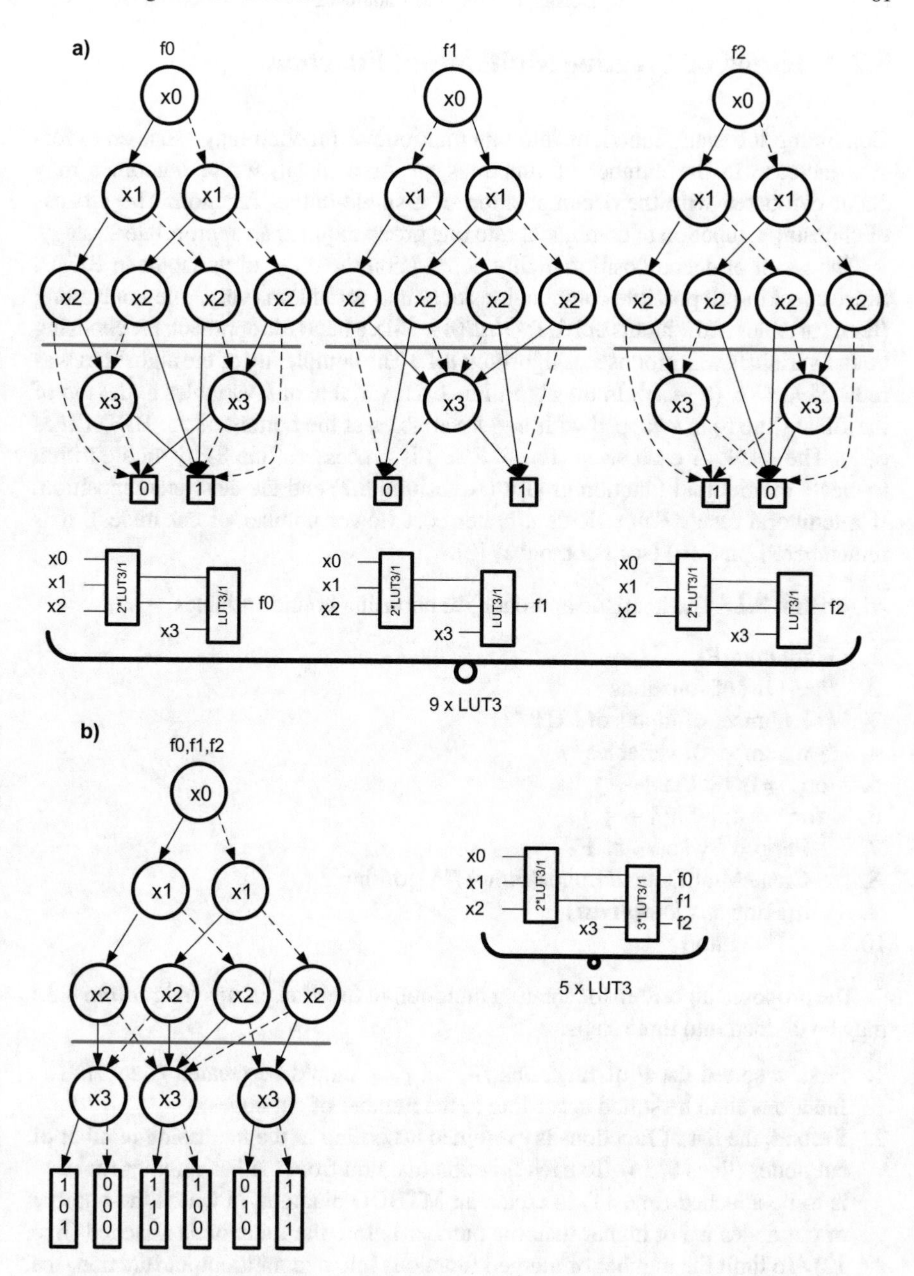

Fig. 8.6 Function diagrams and their implementations: **a** single functions (separated bound blocks), and **b** multioutput function f_0, f_1, and f_2 (using a common bound block)[2] [7]

---

[2](© (2020) IEEE. Reprinted with permission from Opara et al. [7]).

8.2 Method of Creating Multioutput Function

Combining too many functions into one multioutput function may cause an excessive increase in the number of functions g_i. As a result, worse outcomes may occur compared with the decomposition of a single-output function. The process of choosing a function to combine it into one group requires an appropriate strategy.

The result of decomposition highly depends on the order of variables in BDDs. Since checking all possible combinations of bound variables is very time-consuming (for n variables and k inputs of LUT: $O(n!/(n-k)!)$, a heuristic approach for choosing bound variables was proposed (Algorithm 8.1.) The complexity of the algorithm was reduced to $O(k*(n-k))$. In the algorithm, each variable of k variables at the top of the BDD (line 5) is swapped with $n-k$ variables at the bottom of the BDD (lines 6, 7). The result of each swap (line 7 *Ftmp*) is processed (line 8) by an algorithm to create multioutput function groups (Algorithm 8.2) and the next decomposition. If a temporal result *Ftmp* yields a better cost (lower number of cut nodes), it is remembered (line 10) (see Footnote 1) [2].

Algorithm 8.1: The heuristic approach for choosing bound variables

```
1.  FindOrder(F)
2.  /* F[] list of functions */
3.  /* k number of inputs of LUT */
4.  /* n number of variables */
5.  for(i = 0; i < k; i + + )
6.    for(j = k; j < n; j + + )
7.      Ftmp = Swap(xi, xj, F)
8.      CreateMultiOutputGroups(Ftmp) //Algorithm 2
9.      if(Ftmp has lower cost)
10.        F = Ftmp
```

The proposed algorithm for creating multioutput function groups (Algorithm 8.2.) may be divided into three steps:

1. First, a sorted list F of functions $f_0, \ldots, f_{m-1}$ should be created (line 5). The functions shall be sorted according to the number of cut nodes.
2. Second, the list of functions is examined according to the increasing number of cut nodes (lines 8, 16). To each function diagram from the list, another function is to be attached (line 11) to create an MTBDD diagram, in which the number of cut nodes is not higher than the number before the function is attached (line 12). To limit the number of merged functions into one multioutput function, the parameter α is employed (line 10).
3. Third, the last step is based on assigning the codes to cut nodes and performing the function g_i in a bound block (line 17) (see Footnote 1) [2].

Algorithm 8.2: The creating multioutput function groups

```
1.  CreateMultiOutputGroups(F)
```

```
2.  /* F[] list of functions */
3.  /* m number of functions */
4.  /* α max num of merged functions */
5.  Sort(F[])
6.  F[m] = null;
7.  i = 1
8.  while(F[i]! = null)
9.    for(j = i-1; j > = 0; j–)
10.     if(num. merged fun. into F[j] < α)
11.       T = F[i] ∪ F[j] //merge BDDs
12.       if(CutNodes(T) = CutNodes(F[i]))
13.         F[j] = T
14.         Remove(F[i])
15.         i--
16. i++
17. AssignCodes(F[])
```

The algorithm is illustrated by the following example. Figure 8.7 presents a multi-root binary decision diagram (SBDD) of the functions that describe the benchmark

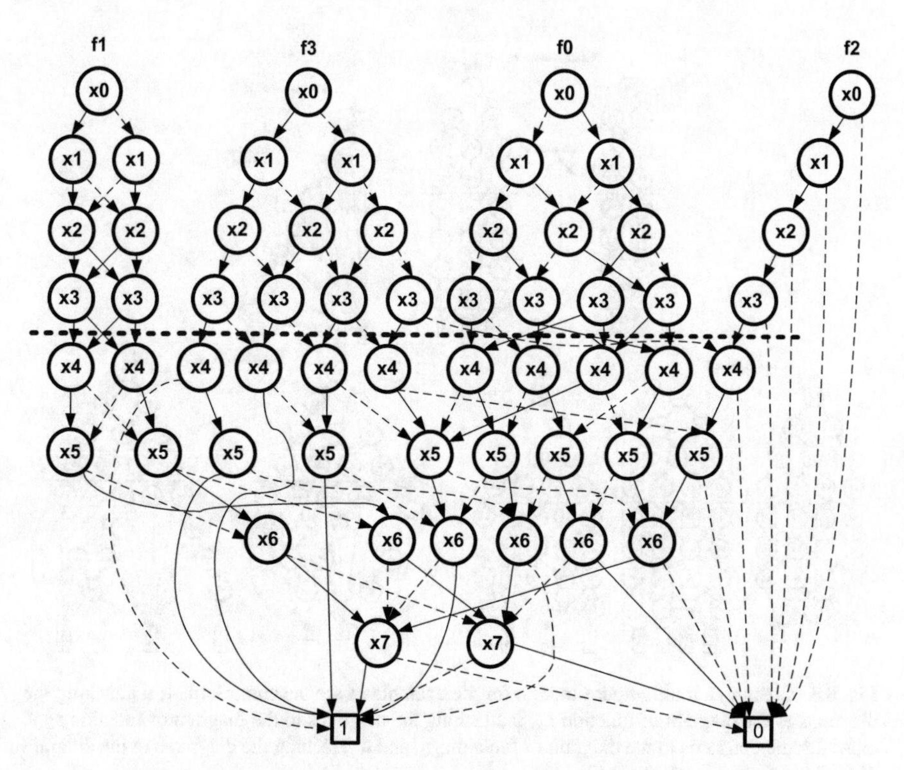

Fig. 8.7 SBDD of rd84 benchmark (see Footnote 1) [2]

rd84. The diagram shows the cutting line that is below the nodes associated with the variable x_3. Four levels of nodes are placed above the cutting line and correspond with the four bound variables x_0, x_1, x_2, and x_3. The number of outputs of a bound block depends on the number of cut nodes below the cutting line. Regarding separate function diagrams, the following number of cut nodes exist: $f_0 - 4, f_1 - 2, f_2 - 2, f_3 - 5$ (see Footnote 1) [2].

In accordance with a given algorithm, the sorted list of functions is created. The list is sorted according to the increasing number of cut nodes: f_1 – 2 nodes, f_2 – 2, f_0 – 4, f_3 – 5. Each function is attached to another function to create an MTBDD diagram, whose number of cut nodes is not higher than that before attaching. The given diagram of a function may be only attached to the diagrams that have the same number of cut nodes or lower number of cut nodes. For instance, the function f_1, which has two cut nodes, may be combined only with the function f_2. After attaching, a diagram with three cut nodes (Fig. 8.8a) is created—this kind of combination is not acceptable. While considering attaching the diagrams to the diagram f_2, the same diagram (Fig. 8.8b) is obtained, as shown in Fig. 8.8a. The next function from the list is f_0 (four cut nodes). The functions f_1 or f_2 may be attached to the diagram f_0. After attaching f_0 and f_1, a diagram that has the same number of cut nodes as the number of cut nodes before attaching is created (Fig. 8.8a)—this kind of connection

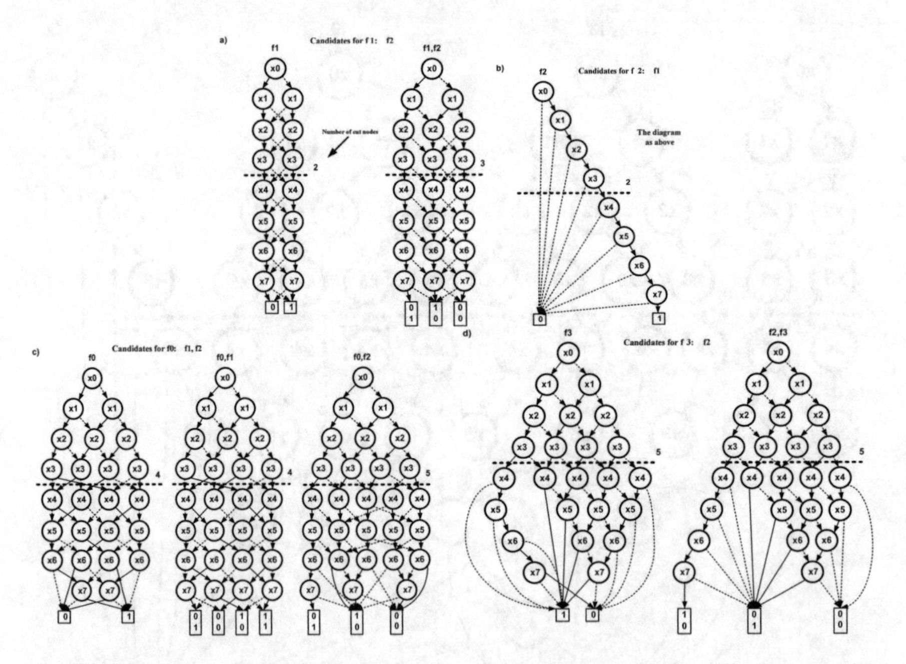

Fig. 8.8 Creating a multioutput function for the example of the benchmark rd84; **a** attaching the diagrams to the diagram of function f_1, **b** attaching the diagrams to the diagram of function f_2, **c** attaching the diagrams to the diagram of function f_0, and **d** attaching the diagrams to the diagram of function f_3 (see Footnote 1) [2]

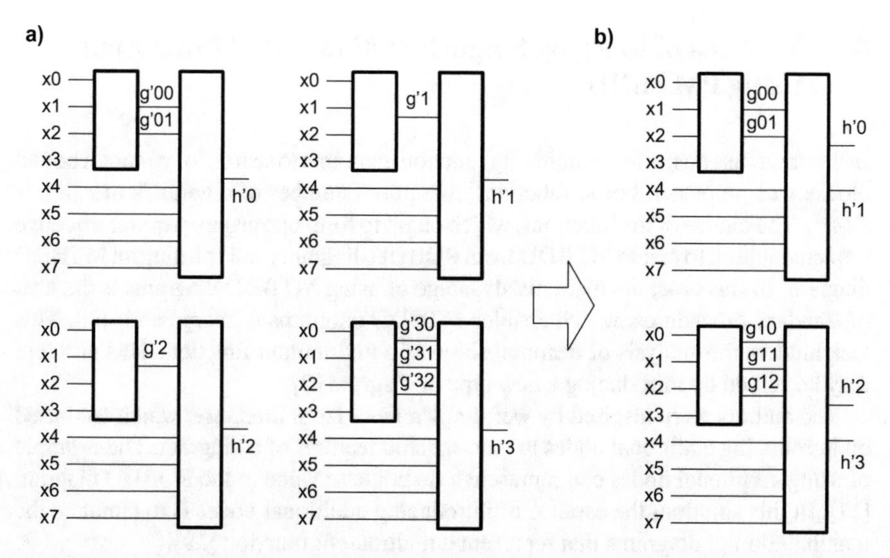

Fig. 8.9 Comparison of a hardware implementation for decomposition of **a** single-output functions and **b** multioutput functions (see Footnote 1) [2]

is acceptable and is the same as the connection in the case of the diagrams f_3 and f_2 (Fig. 8.8d) (see Footnote 1) [2].

The last step of the algorithm is assigning codes to cut nodes and performing the function g_i in a bound block (see Footnote 1) [2].

For the multioutput function f_0, f_1 (four cut nodes), the two coding functions g_{00}, g_{01} are needed. For the multioutput function f_2, f_3, five cut nodes exist. Thus, the three functions g_{10}, g_{11}, g_{12} are needed.

Figure 8.9 presents the comparison of a hardware implementation after decomposition of a multioutput function and a single-output function.

Combining too many functions may cause a substantial and undesirable increase in the number of functions g_i. In the case of a given example, the limitation of the number of multioutput functions to two functions was applied (parameter α in Algorithm 8.2). The lower is the number of functions g_i, the higher is the number of inputs of a free block that is needed. To implement a free block, a higher number of LUTs is required. Reserving several LUTs, which are devoted to implement a bound block, may not be profitable in accordance with the increase in the number of inputs of a free block. For example, function f_1 may be decomposed in two ways: using either the function decomposition $f_1 = h'_1(g'_1, x_4, x_5, x_6, x_7)$ or decomposition of the multioutput function $f_1 = h_1(g_{00}, g_{01}, x_4, x_5, x_6, x_7)$. The function h_1 has one variable more than the function h'_1. To prevent an excessive increase in the number of functions g_i, two strategies may be employed: designate the maximum number of functions in a multioutput function or use a special kind of coding to share some of the functions g_i (discussed in Chap. 9) (see Footnote 1) [2].

8.3 Methods of Merging Single Functions into Multioutput Using PMTBDD

In the previous subsection, combining function diagrams to search for common bound blocks was proposed. For m functions, a required number of diagram's merging is $O(m^2)$. The choice of the functions, which ought to form one group, requires effective implementation to create MTBDD from ROBDD diagrams and splitting an MTBDD diagram. In this case, the main disadvantage of using MTBDD diagrams is the lack of standard procedures, which enables MTBDD diagrams to merge and split. This lack hinders the analysis of decomposition of a multioutput function. This problem may be solved by introducing a new type of diagram [2].

The authors were inspired by well-known ideas from literature, which are based on introducing additional nodes to gain specific features of a diagram. The example of using additional nodes can represent a do not care value in the ROBDD diagram [11]. In this solution, the essence of introducing additional nodes is to simulate the manipulation of diagrams that represent a multioutput function [19].

We introduce the notion of a pseudo MTBDD (PMTBDD) diagram. A PMTBDD diagram is referred to as an MTBDD diagram, in which the leaves were replaced with specific ROBDD subdiagrams. Each sub-diagram consists of additionally introduced variables. Thanks to these variables it will be possible to merge ROBDD diagrams into PMTBDD diagrams utilizing the standard procedure bdd_or() [20]. Each single function that is one output of the function of many outputs requires the introduction of an additional node [2].

Each leaf of the MTBDD diagram of the m multioutput function f_i: $B^n \rightarrow B$, where $i = 0, \ldots, m$-1, holds a vector of values of multioutput function f_i. This vector is replaced by a subdiagram of the following expression:

$$\sum_{i=0}^{m-1} f_i \cdot f_i' \tag{8.1}$$

where f_i' are introduced variables [2].

An example of MTBDD leaf representation for a two-output function $f_0 f_1$ was presented in Fig. 8.10. The two additionally introduced variables are named f_0', f_1'. The vector of multi-output function values corresponds to subdiagrams of the following expressions:

$$11_{f_0 f_1} \rightarrow f_0' + f_1', \quad 10_{f_0 f_1} \rightarrow f_0', \quad 01_{f_0 f_1} \rightarrow f_1', \quad 00_{f_0 f_1} \rightarrow 0 \tag{8.2}$$

More advanced exemplary representations of a PMTBDD diagram for the multi-output function $f_0 f_1 f_2$ is presented in Fig. 8.11. In each case the leaf of an MTBDD diagram placed on the left corresponds to the PMTBDD representation placed on the right. For example, the leaf $111_{f0,f1,f2}$, correlates with a subdiagram of the form $f_0' + f_1' + f_2'$ [2].

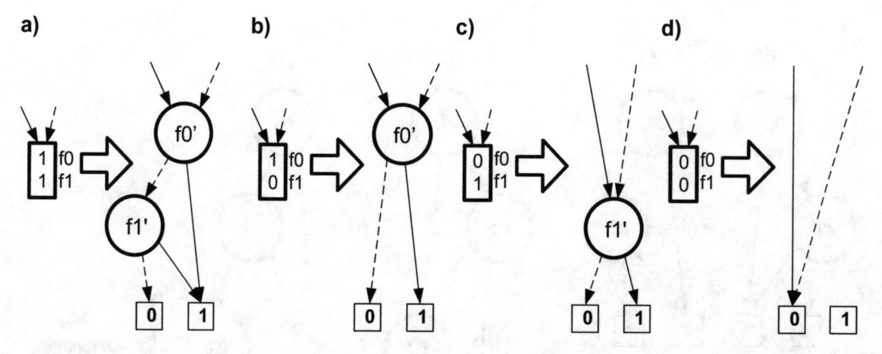

Fig. 8.10 Example of representing MTBDD leaf using ROBDD subdiagram with additional variables: a–d four possible cases of the value of functions f_0, f_1 (see Footnote 2)

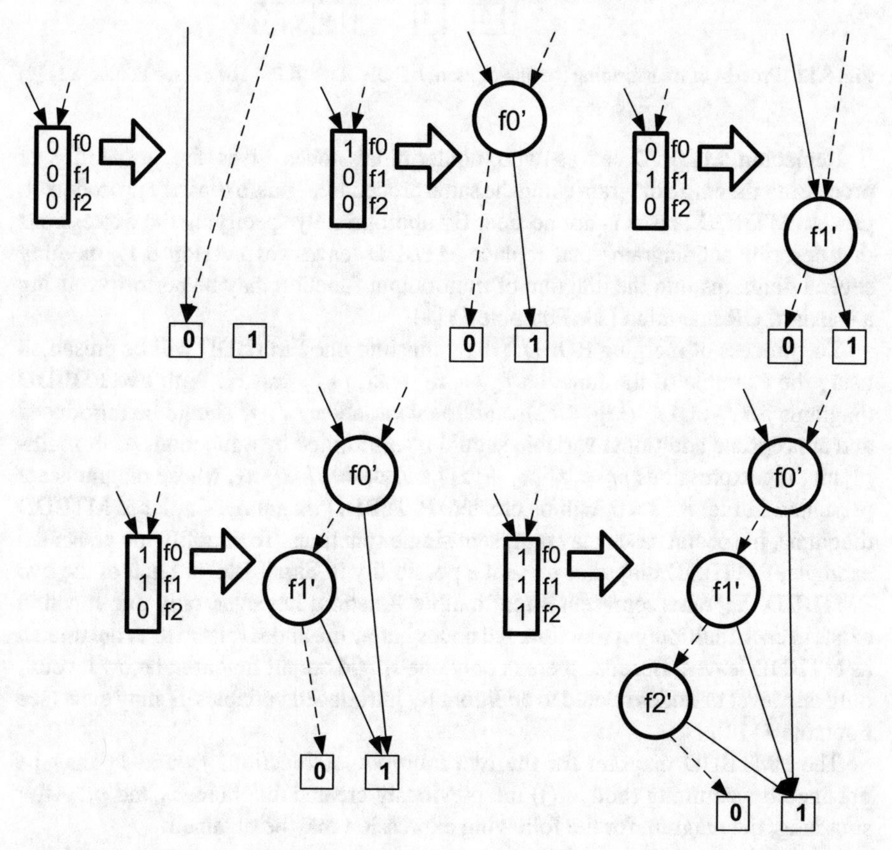

Fig. 8.11 Examples of representations of leaves in the diagrams MTBDD and PMTBDD for a multioutput function f_0, f_1, f_2. (see Footnote 1) [2]

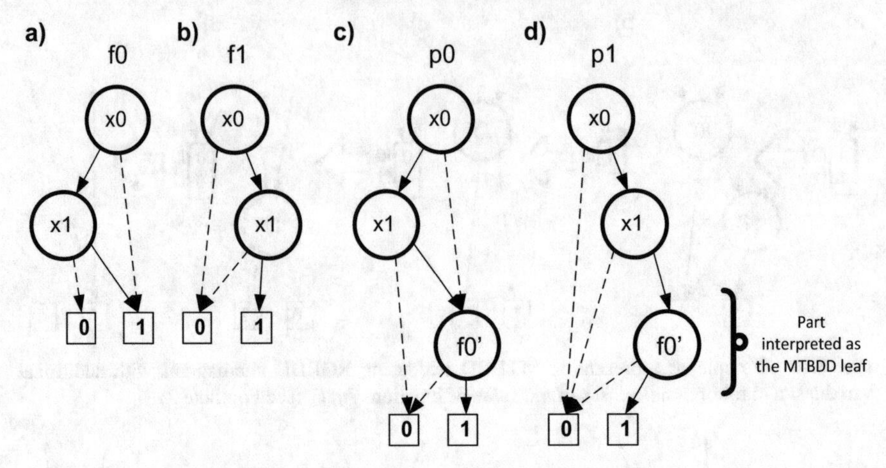

Fig. 8.12 Process of transforming the diagrams: **a**, **b** ROBDD, **c**, **d** PMTBDD (see Footnote 1) [2]

Replacing MTBDD leaves with nonterminal nodes gives the possibility of processing the entire diagram using the same procedure. Thus, a separate procedure to process MTBDD leaves is not needed. By appropriately specifying the expressions that describe subdiagrams that replace MTBDD leaves (expression 8.1), merging several diagrams into the diagram of multioutput function may be performed using a standard OR operation (see Footnote 1) [2].

The process of merging ROBDD diagrams into one PMTBDD will be presented using the example of the function $f_0 = x_0 + x_1, f_1 = x_0 \cdot x_1$. With two ROBDD diagrams for f_0 and f_1 (Fig. 8.12), additional variables $f_0{}'$, f_1' should be introduced and appropriate additional variable should be multiplied by a function. After multiplying, the expressions $p_0 = f_0' \cdot (x_0 + x_1)$ and $p_1 = f_1' \cdot x_0 \cdot x_1$, whose diagrams are presented in Fig. 8.12c, d, will be created. PMTBDD diagrams, similar to MTBDD diagrams, in special cases may represent single functions. To simplify the presented example, PMTBDD diagrams present a possibility in Fig. 8.12c, d. Each of the two PMTBDD diagrams represent a multioutput function; however, only one function exists in each multioutput function. All nodes below the nodes related to x_1 are treated as MTBDD leaves. Because there is only one single output function, below level x_1 only one level of nodes related to additionally introduced variables f_1' may exist (see Footnote 1) [2].

The PMTBDD diagram for the two multioutput functions f_0 and f_1 may be obtained by summing (bdd_or()) the previously created diagrams p_0 and p_1. After summing, the diagram for the following expression may be obtained:

$$p_0 + p_1 = f_0' \cdot f_0 + f_1' \cdot f_1 = \left(f_0' + f_1'\right) \cdot x_0 \cdot x_1 + \left(f_0'\right) \cdot \left(x_0 \cdot \overline{x_1} + \overline{x_0} \cdot x_1\right) \quad (8.3)$$

PMTBDD diagram with corresponding MTBDD diagram is shown in Fig. 8.13 (see Footnote 1) [2].

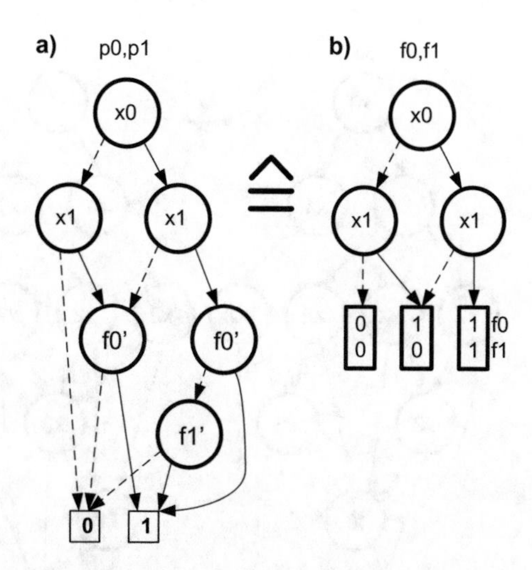

Fig. 8.13 Equivalent forms of the diagrams: **a** PMTBDD, and **b** MTBDD (see Footnote 1) [2]

Consider the case of Fig. 8.14a in which three PMTBDD diagrams represent the functions, respectively: $p_0 = f_0 \cdot f_0'$, $p_1 = f_1 \cdot f_1'$, and $p_2 = f_2 \cdot f_2'$. To merge the diagrams into one PMTBDD logical summing can be applied $p_0 + p_1 + p_2 = f_0 \cdot f_0' + f_1 \cdot f_2' + f_3 \cdot f_3'$. Figure 8.14b and Fig. 8.6b present the symbolic correspondence of PMTBDD and MTBDD diagrams, respectively. The group of nodes, which are described with additional variables, are treated as a leaf. For instance, from the root of MTBDD along the path $0010_{x0,x1,x2,x3}$, the leaf $111_{f0,f1,f2}$ is obtained. Going along the same path in PMTBDD diagram the sub-diagram of is reached. A characteristic feature of PMTBDD diagrams is the ease of the merge and split operations (see Footnote 1) [2, 7].

Splitting is the reverse operation to merging. To create the diagram p0 from the diagram for the set p_0, p_1 setting the variable $f_1' = 0$ is sufficient. Standard procedures (bdd_compose() or bdd_restrict()), which enable substitution of the expression (a constant 0 in this case) with a variable, may be used. As a result, the diagram p_0 shown in Fig. 8.12 is created. Analogously, any function can be removed from the diagram of a multioutput function (see Footnote 1) [2].

The following advantages of the proposed PMTBDD diagrams are listed as follows:

- a separate table of unique identifiers for the leaves is not needed (in the case of MTBDD, while creating a new leaf, its uniqueness needs to be assured to preserve the canonical representation),
- the possibility of rapidly merging the diagrams using a standard sum (bdd_or()) operation on ROBDD (often all two-argument operations are implemented using an ITE operator),
- the possibility of splitting using standard procedures that operate on ROBDD (bdd_restrict() or bdd_compose()) (see Footnote 1) [2].

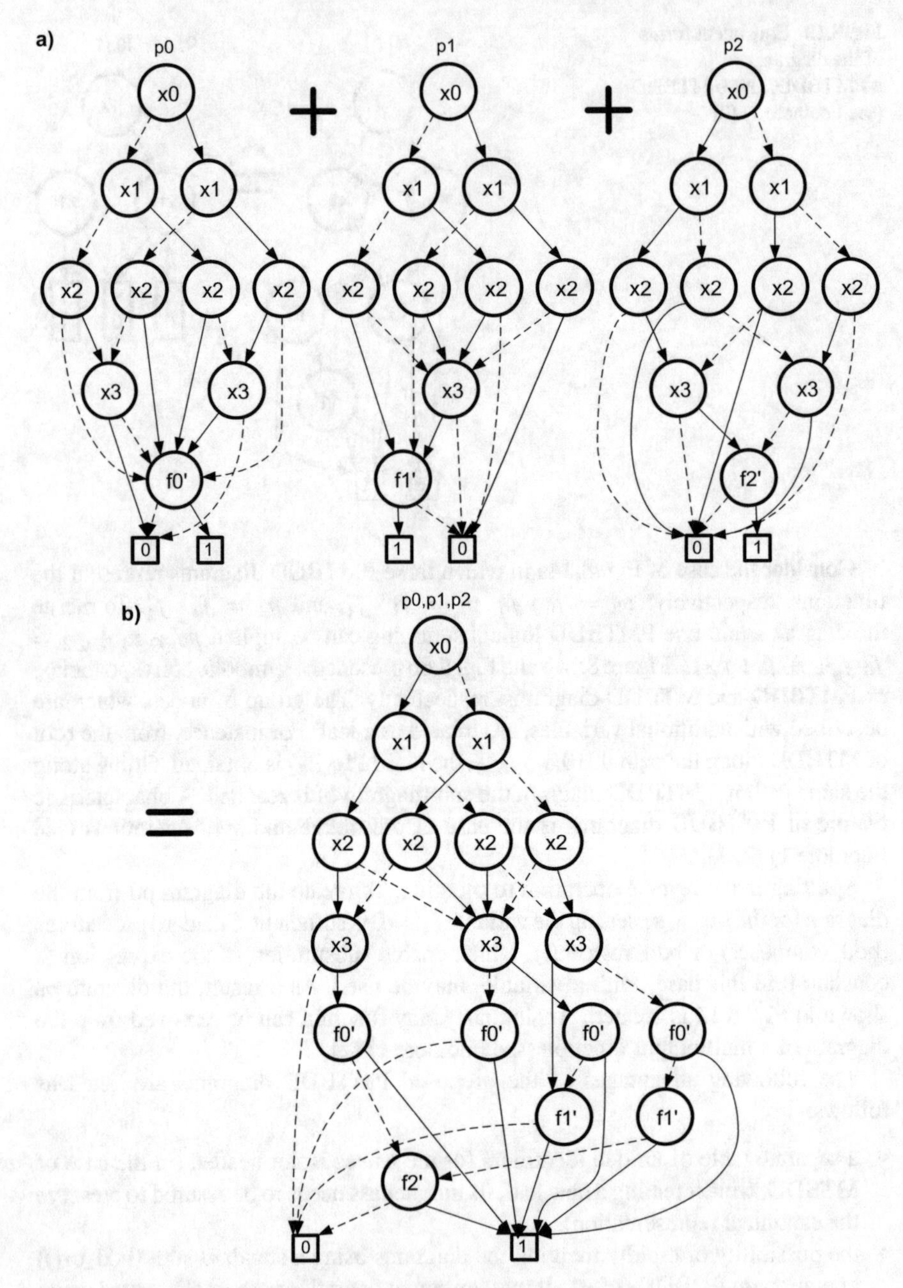

Fig. 8.14 PMTBDD diagrams for the functions from Fig. 8.6a, b (see Footnote 2)

The diagrams that merge may be used in the process of decomposition of multi-output functions. After merging the diagrams, the number of cut nodes does not increase. This notice is the basis of searching for decompositions of appropriate multioutput functions, enabling a gain of common bound blocks (see Footnote 1) [2].

References

1. Kubica M, Kania D, Opara A (2016) Decomposition time effectiveness for various synthesis strategies dedicated to FPGA structures. In: International conference of computational methods in sciences and engineering. American Institute of Physics, Athens, 17 Mar 2016, Seria: AIP Conference Proceedings; vol 1790
2. Kubica M, Opara A, Kania D (2017) Logic synthesis for FPGAs based on cutting of BDD. Microprocess Microsyst 52:173–187
3. Lai Y, Pan KR, Pedram M (1996) OBDD-based function decomposition: algorithms and implementation. IEEE Trans Comput-Aided Des 15(8):977–990
4. Opara A (2006) Wykorzystanie binarnych diagramów decyzyjnych do syntezy wielopoziomowych układow logicznych, Materiały konferencyjne Reprogramowalne Układy Cyfrowe, Pomiary, Automatyka, Kontrola, Szczecin, pp 115–117
5. Opara A, Kania D (2007) Synteza wielowyjściowych układow logicznych prowadząca do wykorzystania wspólnych bloków logicznych, Materiały konferencyjne Reprogramowalne Układy Cyfrowe, Pomiary, Automatyka, Kontrola, Szczecin, pp 39–42
6. Wurth B, Schlichtmann U, Eckl K, Antreich KJ (1999) Functional multiple-output decomposition with application to technology mapping for lookup table-based FPGAs. ACM Trans Des Autom Electr Syst 4(3):313–350
7. Opara A, Kubica M, Kania D (2019) Methods of improving time efficiency of decomposition dedicated at FPGA structures and using BDD in the process of cyber-physical synthesis. IEEE Access 7:20619–20631. doi: 10.1109/ACCESS.2019.2898230
8. Opara A, Kubica M, Kania D (2018) Strategy of logic synthesis using MTBDD dedicated to FPGA. Integration VLSI Journal 62:142–158
9. Opara A, Kubica M (2017) Optimization of synthesis process directed at FPGA circuits with the usage of non-disjoint decomposition. In: Proceedings of the international conference of computational methods in sciences and engineering 2017. American Institute of Physics, Thessaloniki, 2017.04.21, Seria: AIP Conference Proceedings; vol 1906, Art. no. 120004
10. Opara A, Kubica M (2018) The choice of decomposition path taking non-disjoint decomposition into account. In: Proceedings of the international conference of computational methods in sciences and engineering 2018. American Institute of Physics, Thessaloniki, 14 Mar 2018, Seria: AIP Conference Proceedings; vol 2040, Art. no. 080010
11. Minato S (1996) Binary decision diagrams and applications for VLSI CAD. Kluwer Academic Publishers, Boston
12. Kubica M, Kania D (2015) SMTBDD: new concept of graph for function decomposition, IFAC conference on programmable devices and embedded systems PDeS, Cracow vol VII, 519, pp 61–66
13. Kubica M, Kania D (2016) SMTBDD: new form of BDD for logic synthesis. Int J Electron Telecommun 62(1):33–41
14. Kubica M, Kania D (2017) Area-oriented technology mapping for LUT-based logic blocks. Int J Appl Math Comput Sci 27(1):207–222
15. Kubica M, Kania D (2017) Decomposition of multi-output functions oriented to configurability of logic blocks. Bull Polish Acad Sci Tech Sci 65(3):317–331

16. Kubica M, Kania D (2019) Technology mapping oriented to adaptive logic modules. Bull Polish Acad Sci Tech Sci 67(5):947–956
17. Kubica M (2014) Decomposition and technology mapping based on BDD—PhD thesis, Gliwice. https://ssuise-keit.multimedia.edu.pl/doktoraty.php (in Polish)
18. Opara A (2009) Decompositional methods of combinational systems synthesis based on BDD—PhD thesis, Gliwice. https://ssuise-keit.multimedia.edu.pl/doktoraty.php (in Polish)
19. Opara A, Kubica M (2016) Decomposition synthesis strategy directed to FPGA with special MTBDD representation. In: International conference of computational methods in sciences and engineering. American Instytut of Physics, Athens, 17 Mar 2016, Seria: AIP Conference Proceedings; vol 1790
20. Bryant RE (1986) Graph based algorithms for Boolean function manipulation. IEEE Trans Comput C-35(8):677–691

Chapter 9
Partial Sharing of Logic Resources

Decomposition of multioutput functions naturally offers the possibility of sharing some parts of the logic resources [1]. The effective sharing of resources limits the number of logic blocks that are needed to implement the multioutput function. The most basic aspect of the sharing of logic resources is a search for common bound functions, which enables the implementation, in which one bound block is attached to several free blocks that are connected with separate functions and that belong to the set[1] [2].

The sharing of the bound functions g_i may be either full [3] or partial [4], as illustrated in Fig. 9.1.

The basis of the full sharing of bound functions is that all bound block outputs p form the input for all free blocks. In the case of a partial sharing of bound functions, only certain chosen outputs are common. sharing of all bound functions may cause an excessive increase in the number of logic blocks. Thus, partial sharing is a more effective solution. The essence of partial sharing of the bound functions g is the assignment of appropriately chosen codes to the cut nodes of the diagrams of separate multioutput functions. The concept of assigning appropriate codes is based on unicoding [5]; it is the process of assigning only one code to each node. One way to create the codes assigned to the cut nodes is to form all possible bound functions g_i for single-output functions in explicit [6] or implicit [7] form. Scholl [8] presents a method for directly obtaining common bound functions g without the need to analyze all options. To determine a general algorithm for finding common bound functions, we consider the first example (see Footnote 1) [2].

Example 9.1 Figure 9.2 presents decomposition tables for the functions f_0 and f_1. The variables x_0, x_1, and x_2 create a bound set. Each table has four column patterns (f_0 – a, b, c, d, f_1 – k, l, m, n) that correspond to the four cut nodes in the diagram (Fig. 9.3a, b). To implement the bound blocks of each function, two LUTs with three inputs are required. When combining f_0 and f_1 into a multioutput function, six column patterns and six cut nodes (Fig. 9.3c) are required. To distinguish these six

---

[1]

M. Kubica et al., *Technology Mapping for LUT-Based FPGA*, Lecture Notes in Electrical Engineering 713, https://doi.org/10.1007/978-3-030-60488-2_9

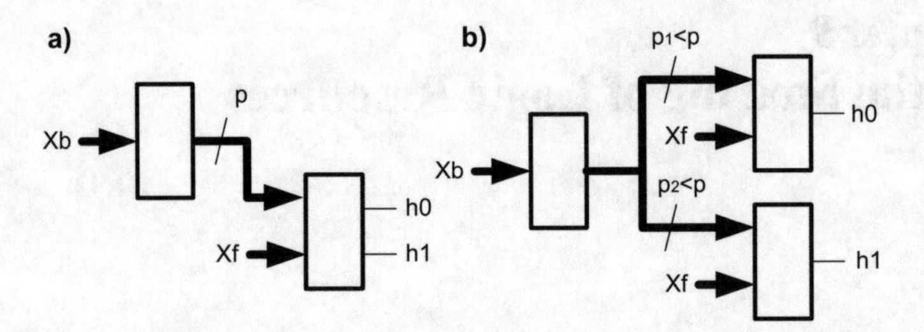

Fig. 9.1 Sharing of bound functions: **a** all functions (full sharing) and **b** only some (partial sharing) (see Footnote 1) [2]

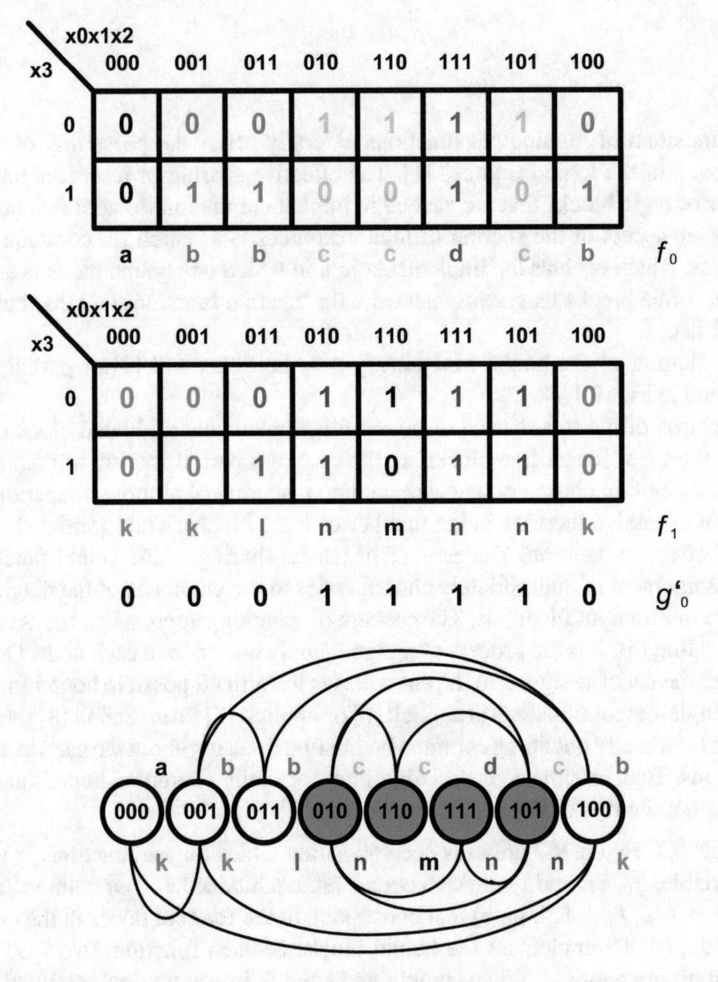

Fig. 9.2 Decomposition tables and pattern compatibility graph for example functions (see Footnote 1) [2]

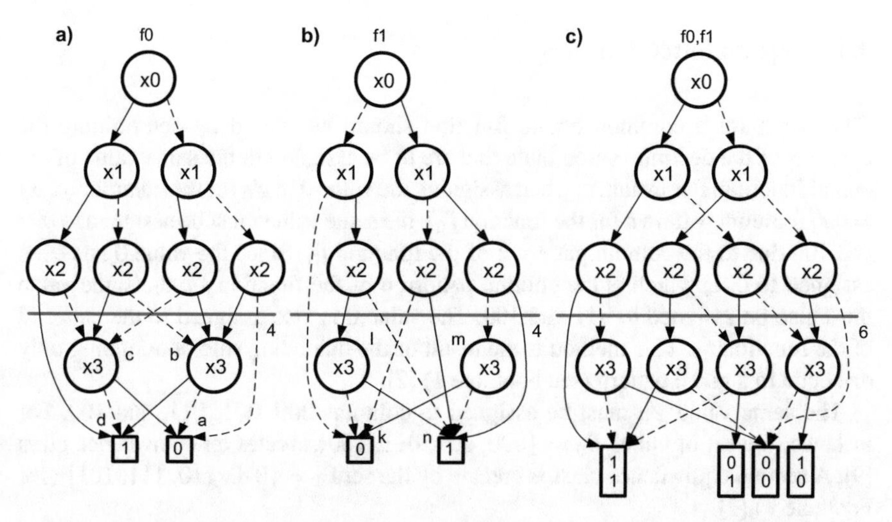

Fig. 9.3 Diagrams of functions with one common bound function (see Footnote 1) [2]

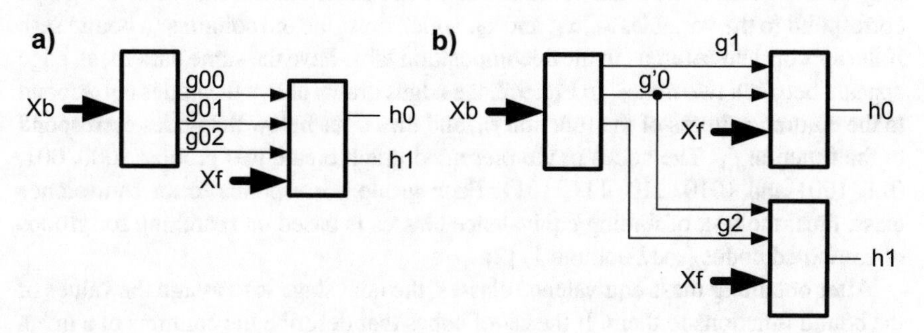

Fig. 9.4 Example of sharing of a function of a bound block: **a** full sharing; and **b** partial sharing (see Footnote 1) [2]

nodes, three bits and three bound functions g_i (Fig. 9.4a) are necessary. Because the cardinality of this bound set is three, this decomposition is pointless from a circuit synthesis point of view because the number of inputs of a free block does not decrease compared with the number of variables of the decomposed function. When using the common bound function g'_0 (Fig. 9.4b), to implement a bound block, three LUTs are needed. However, in terms of free blocks, only two signals are attached from a bound block. In the next part of this section, a method of searching for a bound function will be presented (see Footnote 1) [2].

9.1 Equivalence Classes

The search for a common bound function should be started by determining the columns of the decomposition table that are to be assigned to the same value of the bound function. For instance, when assigning the value 0 to g'_0 for the column $x_0x_1x_2$ = 000 (column pattern a for the function f_0), the same value must be assigned to 001 and 100 due to the column pattern k of the function f_0. Since the value 0 has been assigned to 001, which is the column pattern b of the function f_0, the same value must also be assigned to 011 and 100. The value 0 is also assigned to the pattern l of the function f_1. This method is the result of the unicoding rule of assigning only one code to a given pattern (see Footnote 1) [2].

The same value g'_0 must be assigned to columns 000, 001, 011, and 100. For instance, the set of cubes $E_0 = \{000, 001, 011, 100\}$ creates an equivalence class [9]. A second equivalence class is created by the set $E_1 = \{010, 110, 111, 101\}$ (see Footnote 1) [2].

The pattern compatibility graph shown in Fig. 9.2 illustrates the columns that may be assigned to the same value of the bound function. The nodes of the graph correspond to the variables x_0, x_1, and x_2, which describe the columns (a bound set). If the appropriate columns in the decomposition table have the same pattern, an edge appears between two nodes. In Fig. 9.2, the edges drawn above the nodes correspond to the column patterns of the function f_0, and the edges below the nodes correspond to the function f_1. The nodes in the presented graph create two groups: {000, 001, 011, 100} and {010, 110, 111, 101}. Each group corresponds to an equivalence class; thus, the task of finding equivalence classes is based on searching for groups of combined nodes (see Footnote 1) [2].

After obtaining these equivalence classes, the next stage is to assign the values of the bound functions to them. If the set of cubes that describe the columns of a given pattern ($K_b = \{001, 011, 100\}$ for the pattern b) is included in a given equivalence class ($K_b \subset E_0$), this pattern belongs to an equivalence class. The number of common bound functions depends on the equivalence classes, and the maximum number of patterns belong to one class. In the case where only one equivalence class is identified, no common bound function will exist, because this function must be constant (all patterns belong to one class). When the number of equivalence classes is n_e, the number of common bound functions w for a multioutput function (m) may be limited using the following conditions:

$$\begin{aligned} w &\leq \lceil \log_2(n_e) \rceil, \\ w &\leq \lceil \log_2(\nu(X_f|X_b, f_i)) \rceil, \ i = 0, \ldots, m-1 \end{aligned} \tag{9.1}$$

These conditions only determine the maximum number of bound functions; they do not help to create the truth table for these functions. For the analyzed example, we have $n_e = 2$, $\nu(X_f \mid X_b, f_0) = \nu(X_f \mid X_b, f_1) = 4$. Thus, two bound functions exist for each function but only one function may be common. One bit code, which is generated by the common bound function g'_0, enables us to differentiate two

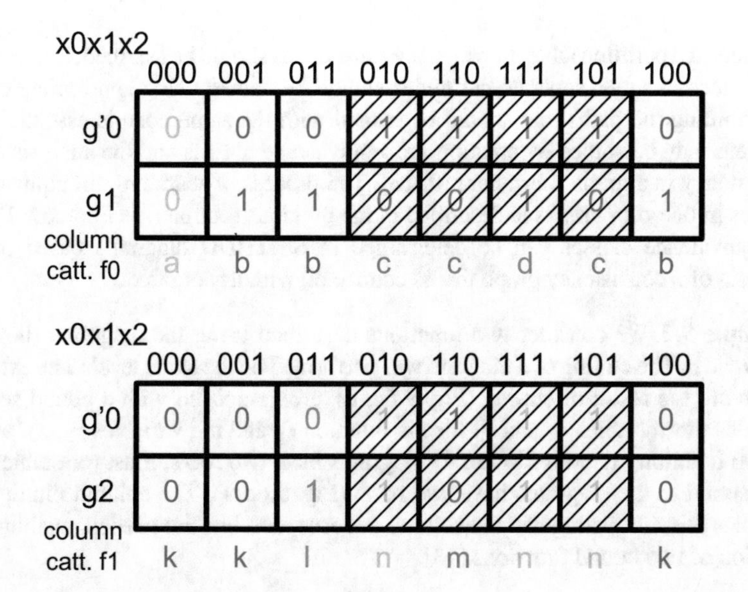

Fig. 9.5 Accepted method of coding (see Footnote 1) [2]

classes—E_0 and E_1. The second function of each pair of bound functions must enable us to differentiate column patterns that belong to one equivalence class. Therefore, no more than two column patterns can belong to each equivalence class for each function. This condition is fulfilled in the case of the analyzed example f_0: $E_0 = K_\mathbf{a} \cup K_\mathbf{b}$, $E_1 = K_\mathbf{c} \cup K_\mathbf{d}$, f_1: $E_0 = K_\mathbf{k} \cup K_\mathbf{l}$, $E_1 = K_\mathbf{m} \cup K_\mathbf{n}$. The accepted method of coding is illustrated in Fig. 9.5 (see Footnote 1) [2].

Equivalence classes for a multioutput function can be created by searching for groups of combined nodes in a graph, as shown in Fig. 9.2. This graph does not have to be explicitly created. In the literature, the algorithm that searches for equivalence classes using ROBDD [9] is described. In this algorithm, the required number of operations on ROBDD linearly depends on the number of edges in the graph, as shown in Fig. 9.2 (see Footnote 1) [2].

9.2 Partial Sharing in SMTBDD

The same process of searching for shared blocks may be conducted for multiroot SMTBDDs [10–14]. The way of proceeding is nearly the same, with the exception that the symbols, which are placed in the cells of a partition table, are connected with an appropriate column pattern in a root table and are not connected to the cut nodes.

The essence of this solution is the appropriate coding of cut nodes, in which only one code is always ascribed to a single node (unicoding). The process of searching for a common bound function is a two-stage process. In the first stage, determining the columns of a partition table, to which the same codes should be ascribed, is necessary. In the second stage, separate cut nodes should be named using appropriate codes. Determining the paths that should be named with the same code is essential. The problem may be solved by creating i.e., equivalence classes and the analysis of the consistency in a graph's structure [9, 15]. The process of searching for equivalence classes in one-root graphs is described in the previous section (see Footnote 1) [2].

Equivalence classes can be determined in SMTBDD diagrams based on the analysis of a consistency graph that is connected with a root table.

Example 9.2 We consider two functions described using the diagrams shown in Fig. 9.6a, b. The cutting of a diagram was performed on the same levels The extracts, which are the result of cutting (Fig. 9.6c, d), are associated with a bound set that includes the variables x_2 and x_3 for the function f_0 and the variables x_2, x_3, and x_4 for the function f_1. Both SMTBDD diagrams have two roots, thus, root tables that correspond to the diagrams have two rows (Fig. 9.6e, f). The column multiplicity of tables is 4. To distinguish the column patterns, two bits are needed, enabling the creation of two bound functions [13].

To find a common coding bit, a consistency graph of a root table, whose nodes will be connected with appropriate combinations of variables in the analyzed extract, shall be created (Fig. 9.7b). Each combination (path) is associated with an appropriate column pattern of a root table. The nodes, which are connected with the same column patterns, can be combined with edges. In the obtained consistency graph of column patterns, the sets that have the edges combined with each other and are marked in appropriate colors in Fig. 9.7b can be distinguished. The sets of consistency graph nodes correspond to equivalence classes. Therefore, three equivalence classes exist. The common bound function g_0 enables the classes to be differentiated in the way that the value 0 corresponds to an equivalence class connected with the set of nodes marked in red, and the value 1 corresponds to the remaining nodes. Basic SMTBDD diagrams in Fig. 9.6c, d may be replaced with the diagrams that include the nodes associated with the bound functions g'_0, g_{10}, and g_{11} (Fig. 9.7c, d) [13].

After placing new SMTBDD diagrams between the cutting lines, ROBDD diagrams, as presented in Fig. 9.8a, b, are obtained. In both diagrams, a node is connected with the common bound function g'_0. The obtained technology mapping is shown in Fig. 9.8c, and common logic resources are marked in red [13].

In the analyzed case, both root tables have the same number of rows. However, in the process of creating a consistency graph, knowing which pattern should be chosen for which combination is necessary instead of what the process is like [13].

9.3 Searching for Equivalence Classes with MTBDD Usage

The previous chapter outlines the use of equivalence classes to find common g functions. This chapter describes how to determine equivalence classes. The proposed

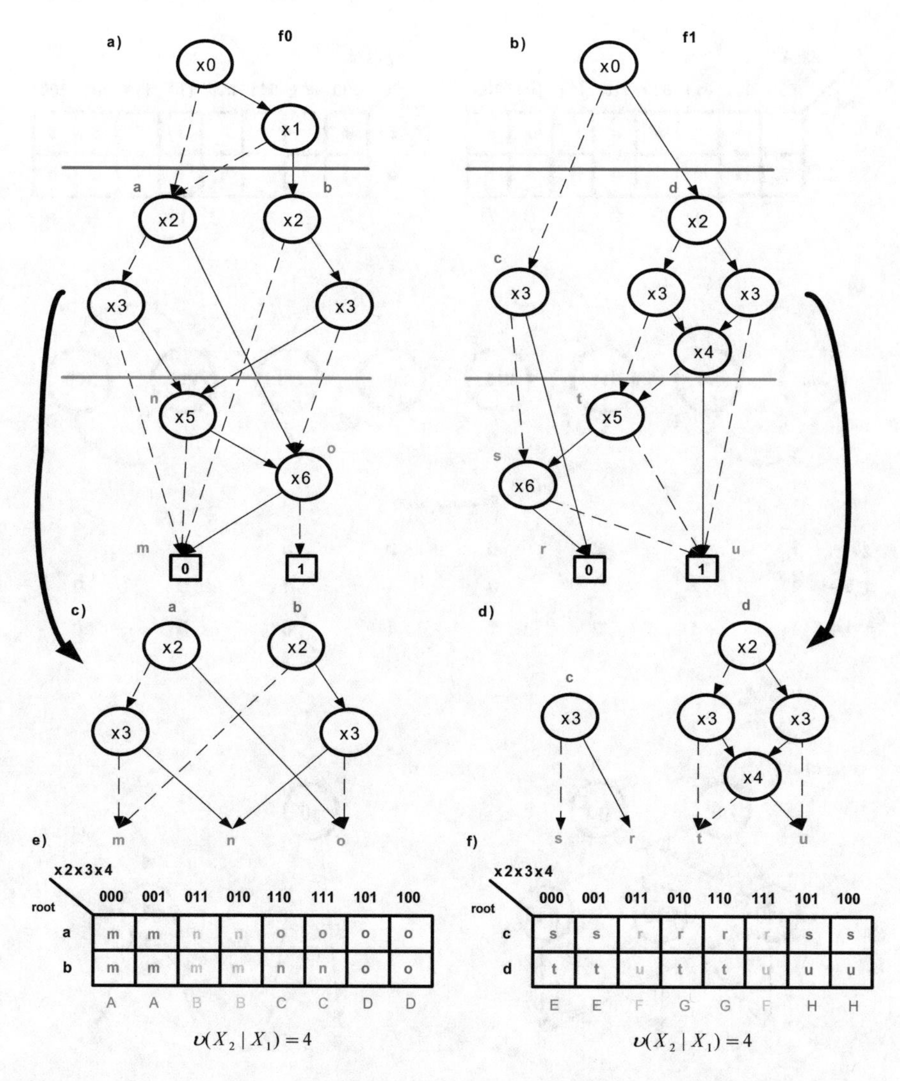

Fig. 9.6 Functions for which equivalence classes are searched: **a**, **b** ROBDD diagrams, **c**, **d** SMTBDD diagrams placed between the cutting lines, and **e**, **f** root tables associated with SMTBDD diagrams

algorithm for searching for equivalence classes uses the process of combining MTBDD diagrams. If the functions are given in the form of ROBDD, then reducing the complexity of searching for equivalence classes to the complexity of performing a single two-argument operation (bdd_apply() [16]) on MTBDD diagrams is possible due to the additional nodes. This algorithm can be thoroughly described using the first example of searching for the equivalence classes of the functions f_0, f_1 (Fig. 9.3). In the description of the algorithm, the following notation is employed: high(**v**) and

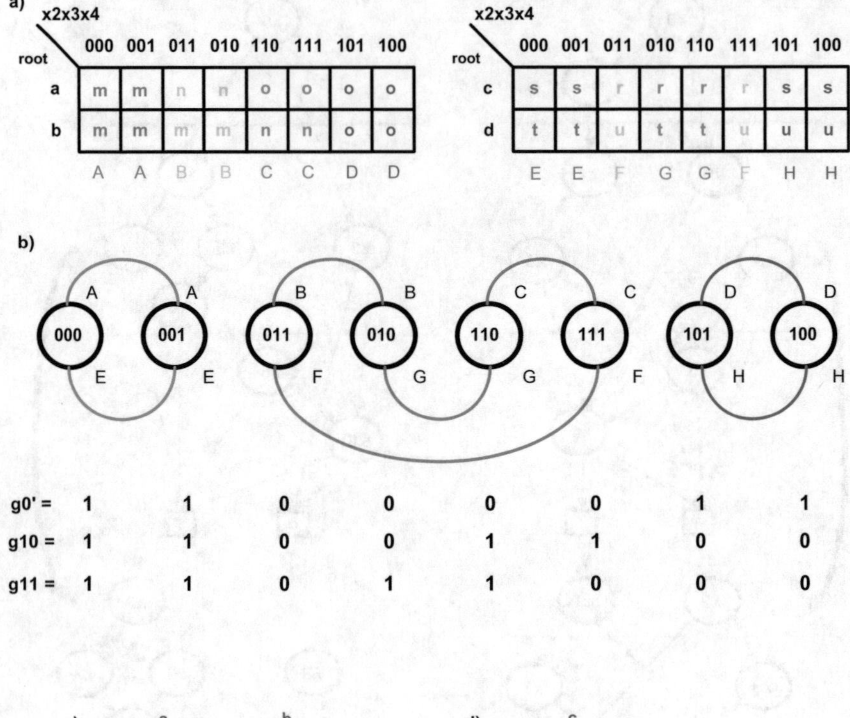

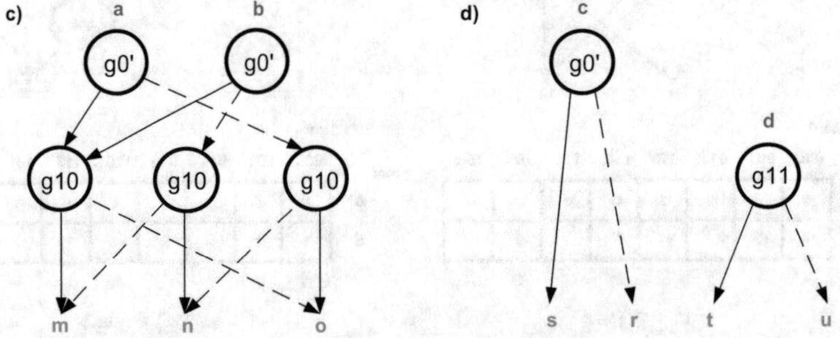

Fig. 9.7 Equivalence classes in SMTBDD: **a** root tables associated with appropriate SMTBDD diagrams, **b** consistency graph with the values of bound functions, and **c**, **d** SMTBDD created after placing the nodes connected with bound functions g (g_0—common function)

low(**v**)—nodes pointed by the "then" and "else" edges of node **v**, index(**v**)—level of the node in a diagram counting from a root node to leaves (Fig. 9.9a). The root node has index 0, the leaf—the largest value (see Footnote 1) [2].

The result of the algorithm will appear in the form of a diagram, in which the number of nodes with edges high(**v**) = 1 will be equal to the number of equivalence classes.

The first stage of this algorithm is based on the appropriate preparation of the function diagrams for further processing. The top parts of the diagram with the

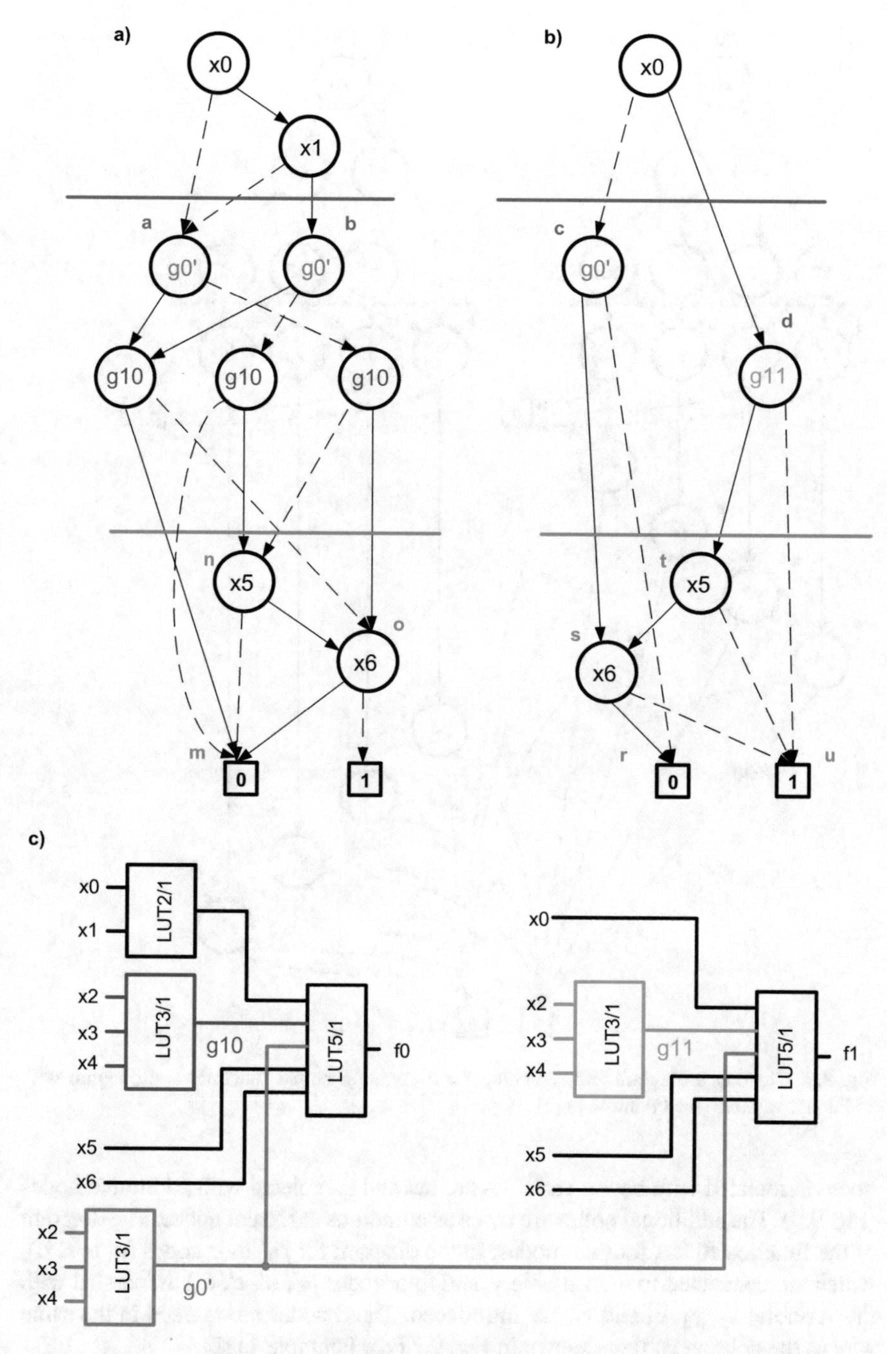

Fig. 9.8 Result of decomposition: **a**, **b** diagrams obtained for common bound functions, and **c** technology mapping in LUT_x/1 blocks

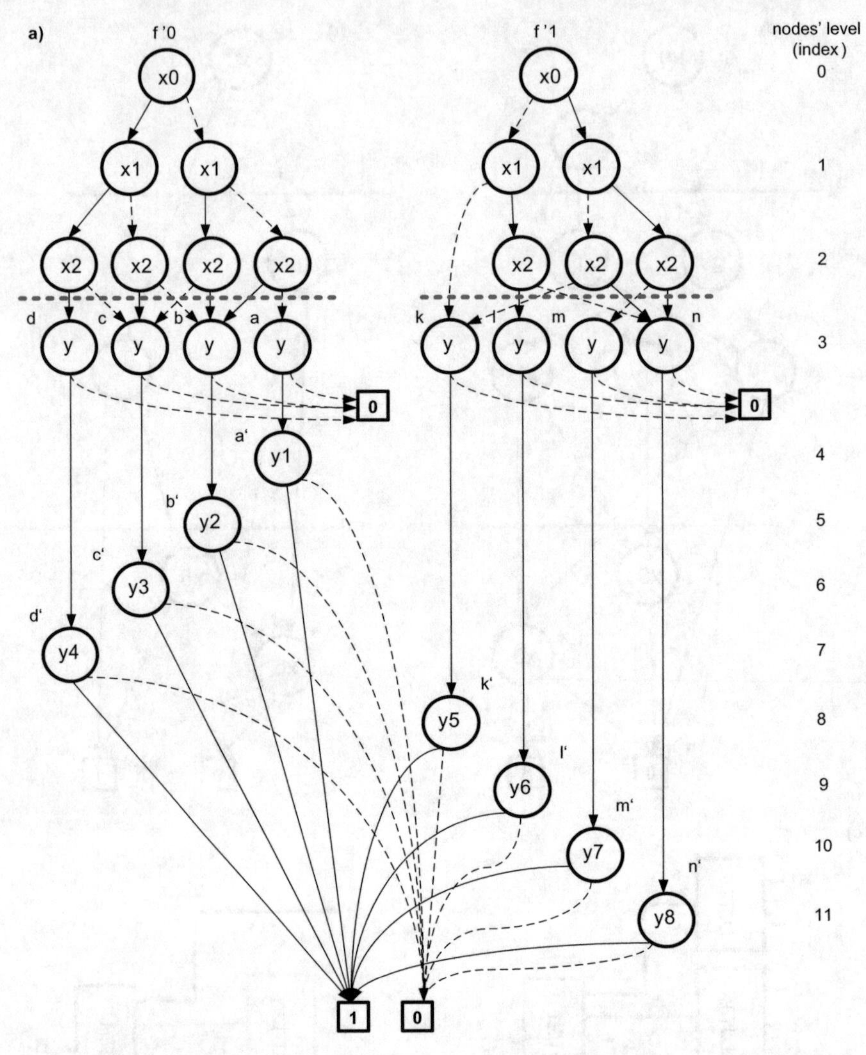

Fig. 9.9 Creating a diagram and searching for a common bound function: **a** a diagram with additional variable (see Footnote 1) [2]

nodes associated with bound variables are cut and completed with additional nodes (Fig. 9.9). The additional nodes are twice as numerous as the cut nodes. The diagram of the function f0 has four cut nodes; in the diagram for f_0' four nodes (**a**, **b**, **c**, **d**), which are connected to the variable y, and four nodes (**a′**, **b′**, **c′**, **d′**) associated with the variables y_1, y_2, y_3 and y_4, are introduced. These nodes are marked in the same way as the column patterns shown in Fig. 9.2 (see Footnote 1) [2].

The additional nodes are treated only as a graph data structure and are not treated as a BDD; therefore, the introduced nodes, whose edge corresponds to the value 0, are not utilized. To make the diagram more legible, these nodes are omitted from

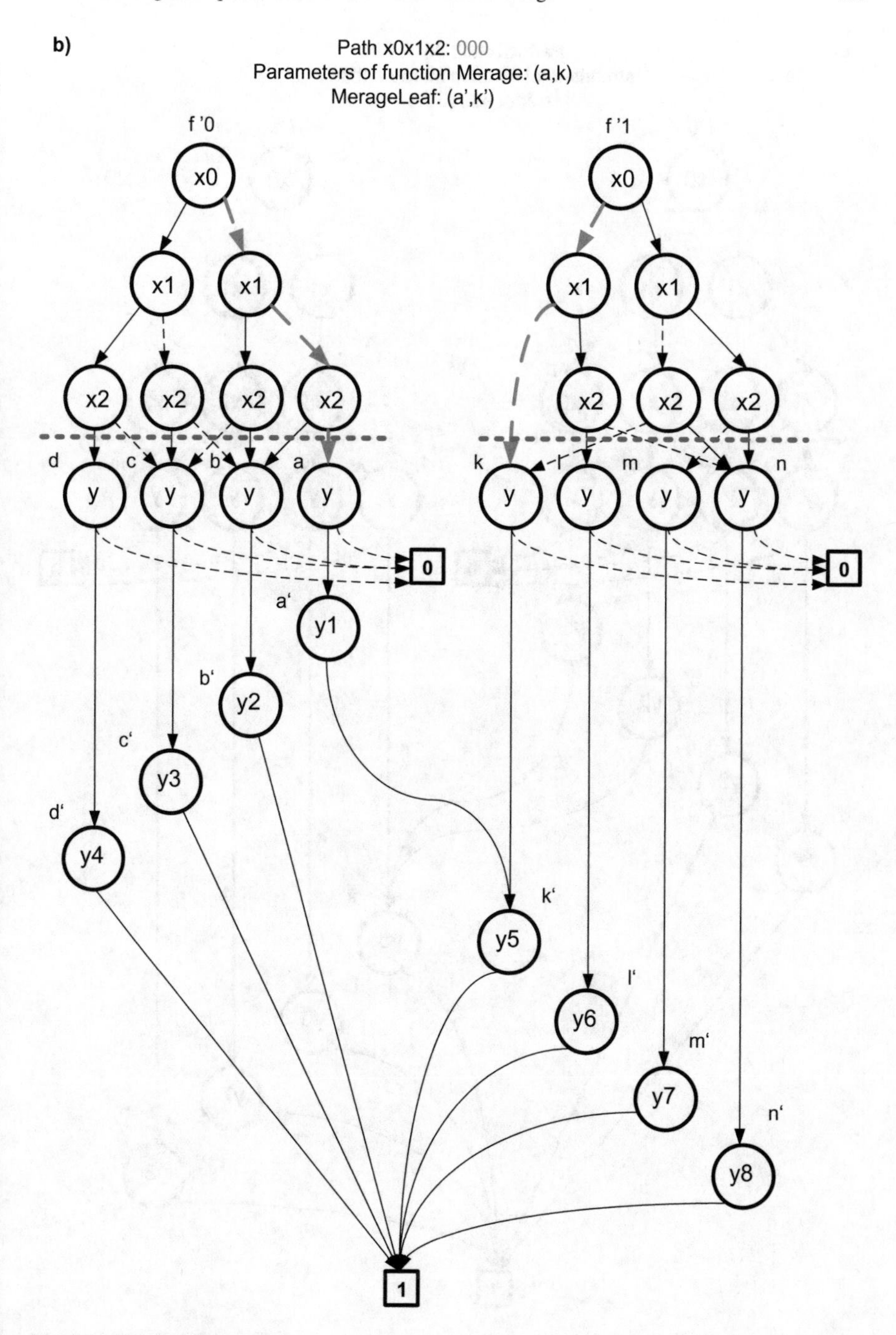

Fig. 9.9 (continued)

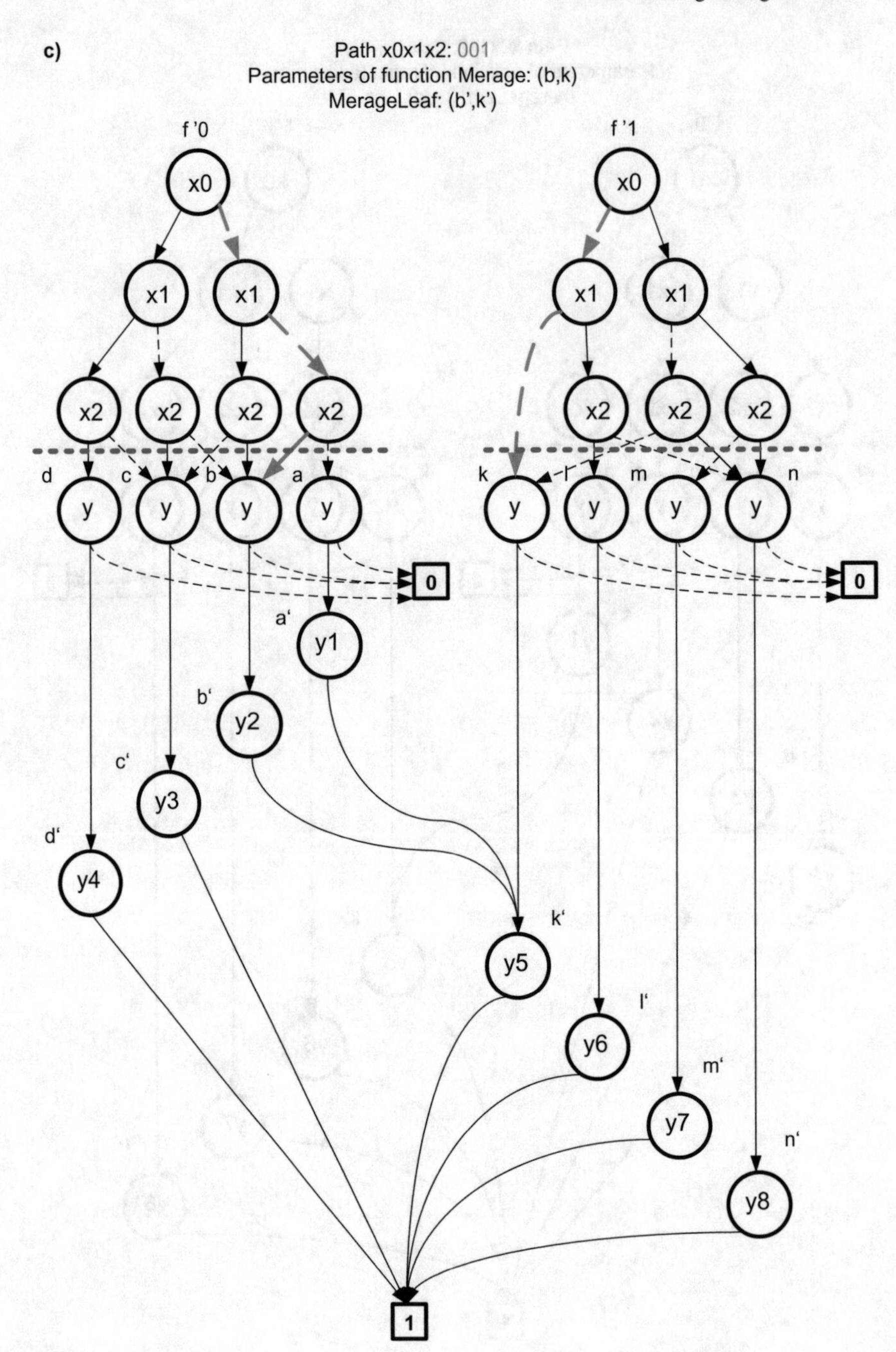

Fig. 9.9 (continued)

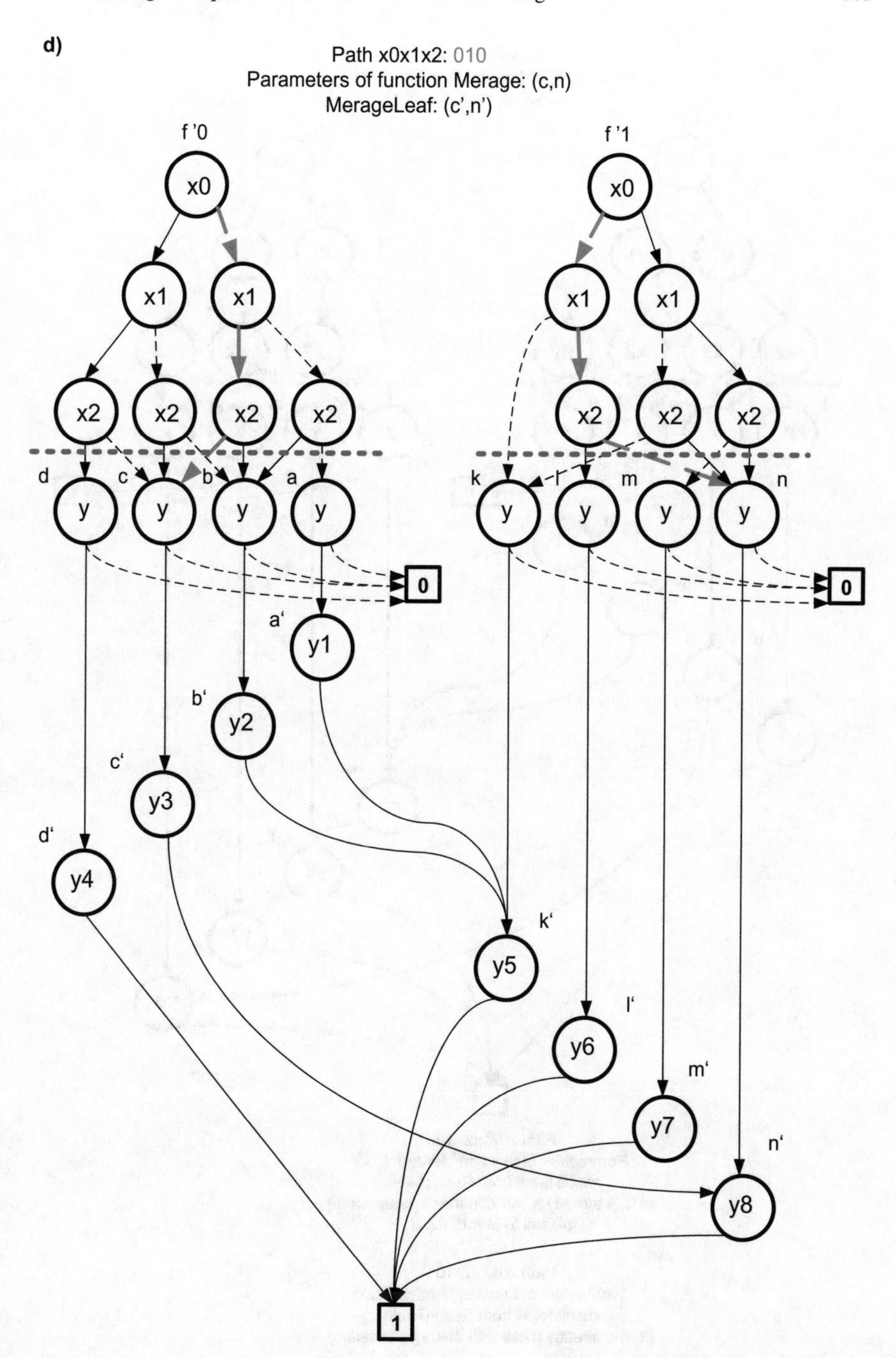

Fig. 9.9 (continued)

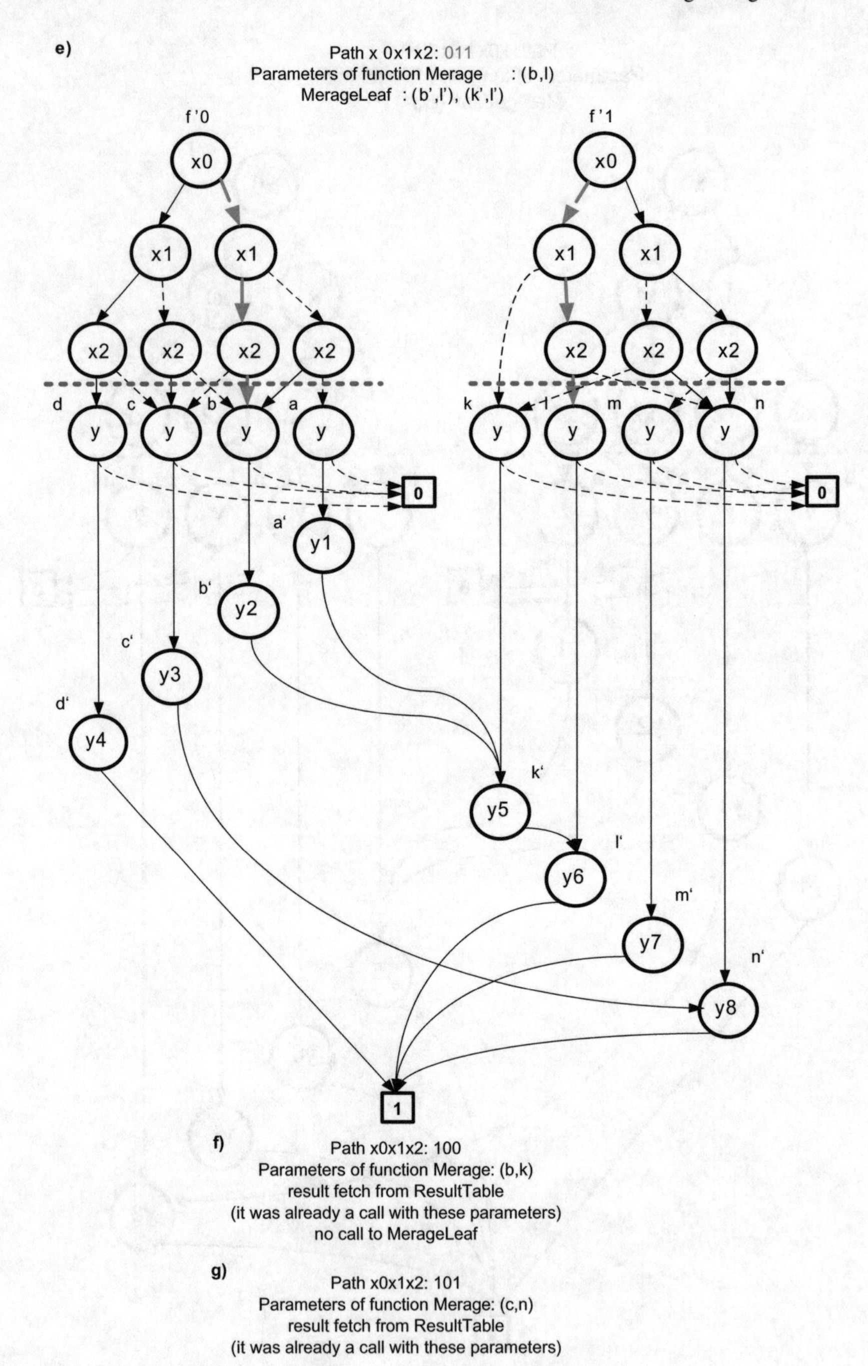
e)
Path x 0x1x2: 011
Parameters of function Merage : (b,l)
MerageLeaf : (b',l'), (k',l')
f'0
f'1
x0
x1
x2
d
c
b
a
k
l
m
n
y
0
a'
y1
b'
y2
c'
y3
d'
y4
k'
y5
l'
y6
m'
y7
n'
y8
1
f)
Path x0x1x2: 100
Parameters of function Merage: (b,k)
result fetch from ResultTable
(it was already a call with these parameters)
no call to MerageLeaf
g)
Path x0x1x2: 101
Parameters of function Merage: (c,n)
result fetch from ResultTable
(it was already a call with these parameters)

Fig. 9.9 (continued)

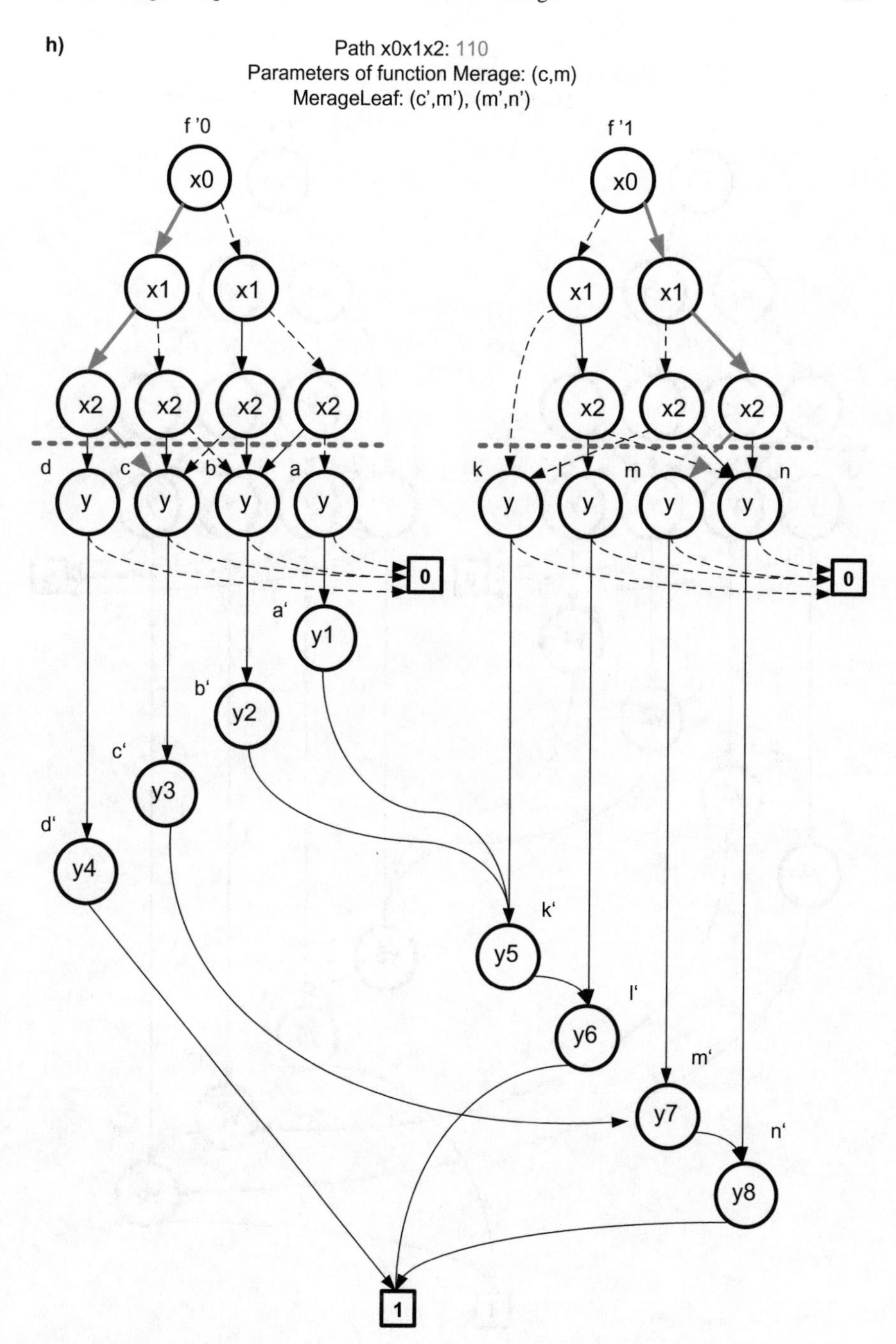

Fig. 9.9 (continued)

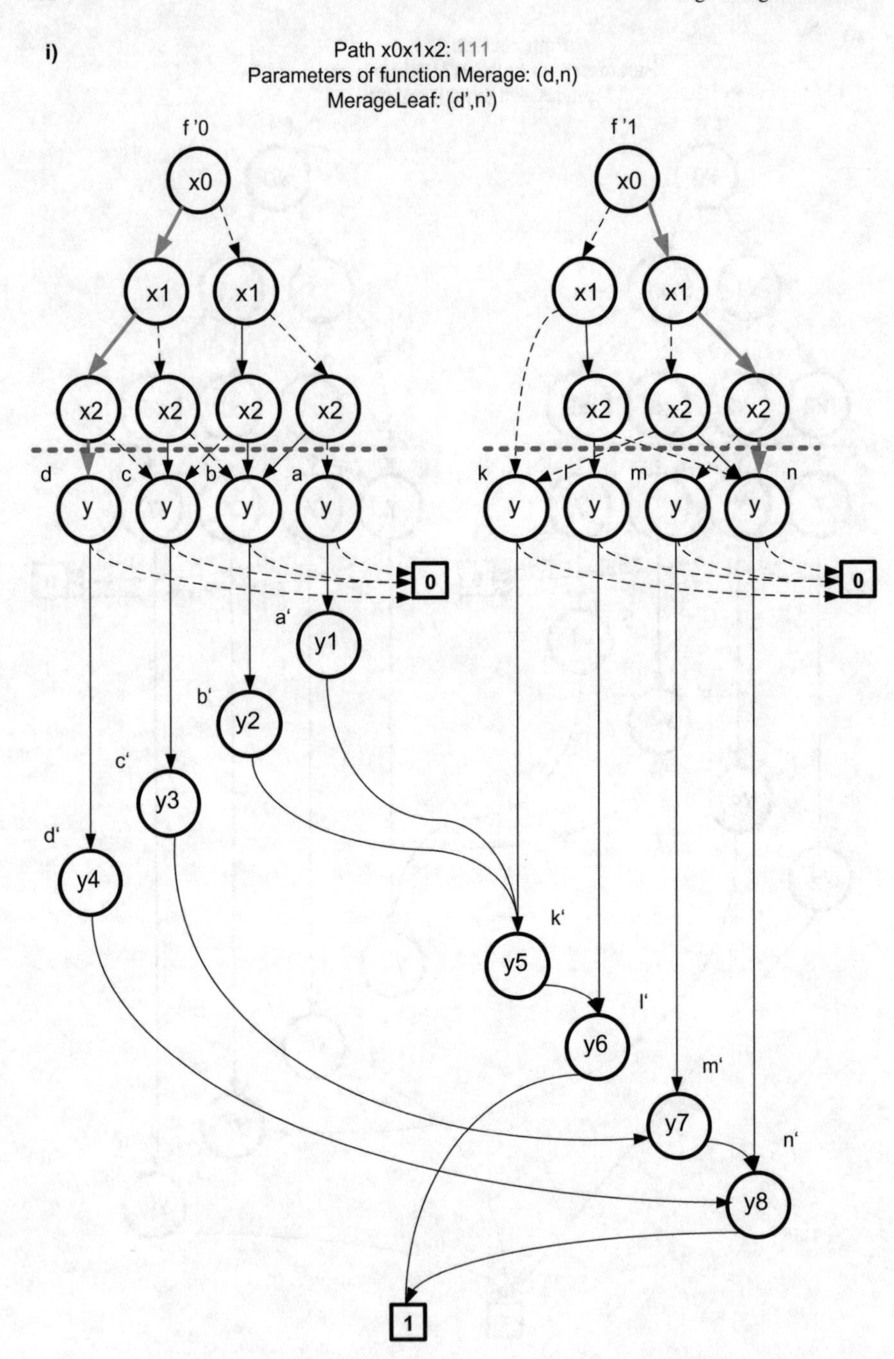

Fig. 9.9 (continued)

Figs. 9.9b–i and 9.10. However, these additionally introduced nodes fulfil only the role of the data structures that are needed to find equivalence classes; they should not be treated as logic expressions (see Footnote 1) [2].

The next stage of the algorithm may be divided into two phases:

- Going through the diagram to the level of additional nodes (Algorithm 9.1, procedure Merge());
- Modification of the edges that connect additional nodes (procedure MergeLeaf()).

The Merge() procedure is similar to the standard procedure bdd_apply() that performs basic two-argument operations that take into account its structure and complexity. One difference is that the Merge() procedure is repeated until the level of the additional nodes is reached, and the bdd_apply() procedure is repeated until the terminal nodes are reached (see Footnote 1) [2].

Algorithm 9.1 Searching for equivalence classes* [2]

```
1.  MergeLeaf(f,g){
2.    /* f,g – The nodes of a function              */
3.    /* connected with additional variables;       */
4.    /* pointers in additional nodes are modified  */
5.    bdd t,hf;   /* Temporary variables  */
6.    if(f ==‘1’ || g ==‘1’)
7.      return;
8.    if(index(f) > index(g)){/*The root has index 0*/
9.      t = f; f = g; g = t;  /* Swap so that f is above g*/
10.   }
11.   hf = high(f);
12.   if(index(high(f)) > index(g))
13.     high(f) = g;
14.   MergeLeaf(hf,g);
15. }
16. Merge(f,g){
17.   /* f,g – Nodes of a multioutput function or a single function with additional
      variables        */
18.   /* n – Number of variables (the index of a leaf)                              */
19.     /*  Diagrams f and g are modified (parts below level n are
      merged)                */
20.   if(f ==0 && g ==0)  return(0);
21.   if(index(f) > n && index(g) > n ){
22.     MergeLeaf(f,g);
23.     return(f);
24.   }
25.   else if( (f,g) ∈ ResultsTable){   return(ResultsTable (f,g));  }
26.   else{
27.     xi = The first variable in the ordering for f or g
```

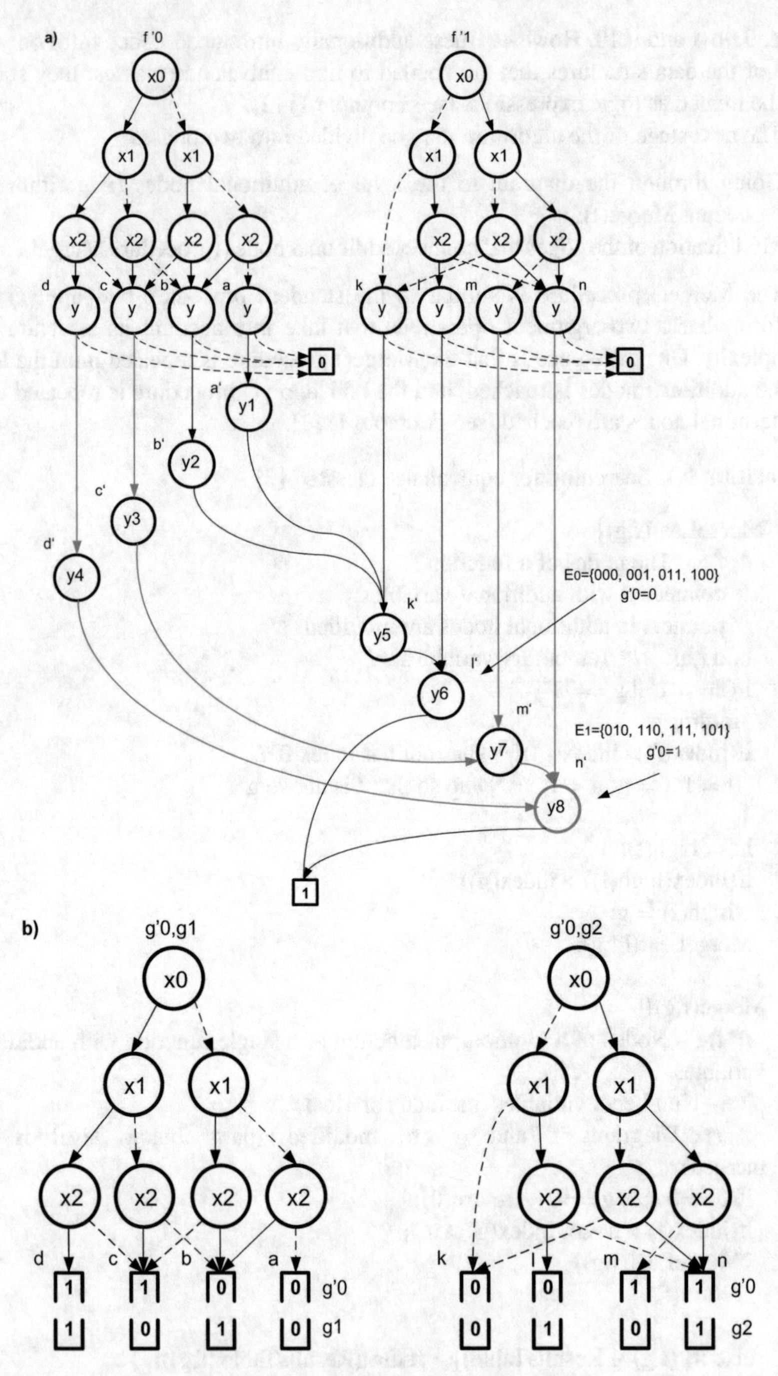

Fig. 9.10 Assignment of codes to bound functions: **a** merged diagrams with two equivalence classes; **b** diagrams of bound functions (see Footnote 1) [2]

```
28.      v = A new node;
29.      low(v) = Merge(f_{xi=0},g_{xi=0});
30.      high(v) = Merge(f_{xi=1},g_{xi=1});
31.      InsertIntoResultsTable(f,g,v);
32.      return(v);
33.    }
34. }
```

In the Merge() procedure, similar to bdd_apply(), the concept is used to store data from previous callings in the ResultsTable. Nodes connected with the variable y are introduced to replace the cut nodes. The cut nodes correspond to the column patterns of a decomposition table. Usage of the ResultsTable until the level of the nodes associated with the variable y is reached enables us to analyze each pair of column patterns only once. As shown in Fig. 9.2, eight columns exist but only six different pairs of patterns ((**a**,**k**), (**b**,**k**), (**b**,**l**), (**c**,**n**), (**c**,**m**), (**d**,**n**)) exist. Due to the use of the ResultsTable, the MergeLeaf() function will be called only six times from the Merge() function. Figure 9.9 presents the diagrams after the following stages of the algorithm. As shown in Fig. 9.9c, for the path $001_{x0,x1,x2}$, the nodes **b** and **k** are attained. For the path 100, the same pair of nodes is also obtained (Fig. 9.9f). However, the MergeLeaf() function is only called for the first time. A similar situation holds in the case of the paths 010 and 101 (Fig. 9.9d and g). This procedure means that a new diagram is not created but the existing diagrams are modified. Thus, the value returned by the Merge() procedure is not important. Although the returned value is written to the ResultsTable, it is only used to record the fact that Merge() was called for a given pair of arguments (see Footnote 1) [2].

The functioning of the MergeLeaf() procedure is based on redirecting the edges of additional nodes in such a way that the edge from the node at a higher level can lead to a node at a lower level. The connection of nodes with this edge adheres to the same equivalence class of appropriate column patterns. For instance, for the path 000 (Fig. 9.9b), nodes **a** and **k** are attained, enabling the calling of MergeLeaf($\mathbf{a}'$, $\mathbf{k}'$) at the next stage. The MergeLeaf() procedure causes a redirection of the edges high($\mathbf{a}'$), which initially indicates leaf 1; thus, it can point to node $\mathbf{k}'$ (line 13, Algorithm 1). The process of merging equivalence classes becomes complicated when the node whose edge has to be redirected leads to a nonterminal node. This situation may arise in the case of the path 011 (Fig. 9.9e) and the calling of MergeLeaf($\mathbf{b}'$, $\mathbf{l}'$). The edge of node $\mathbf{b}'$ points to node $\mathbf{k}'$. Removal of this edge would cause a loss of the information that column patterns **b** and **k** belong to the same equivalence class. Because node $\mathbf{k}'$ is above node $\mathbf{l}'$, edge $\mathbf{b}' \to \mathbf{k}'$ does not change, and the MergeLeaf() procedure is recalled for the second time using the parameters ($\mathbf{k}', \mathbf{l}'$). The three nodes are combined to the chain $\mathbf{b}' \to \mathbf{k}' \to \mathbf{l}'$. The next case, in which the MergeLeaf() procedure has to be repeated several times, occurs when MergeLeaf($\mathbf{c}'$, $\mathbf{m}'$) is called for the path 110 (Fig. 9.9h). The existing edge $\mathbf{c}' \to \mathbf{n}'$ is deleted since node $\mathbf{n}'$ is below node $\mathbf{m}'$. Despite deleting the edge, the information about the equivalence class is not lost. After the next call of the MergeLeaf procedure, the chain $\mathbf{c}' \to \mathbf{m}'$ $\to \mathbf{n}'$ is created (see Footnote 1) [2].

For the example of the transformed diagram, two nodes that have an edge leading to leaf **1** (nodes **l′** and **n′**) were created. These nodes correspond to the two equivalence classes E_0 and E_1 (Fig. 9.10a). The codes were first assigned to the nodes that correspond to the equivalence classes and then assigned to the cut nodes that correspond to the column patterns that belong to these classes. As a result, the diagrams presented in Fig. 9.10b are created (see Footnote 1) [2].

In the process of calling the MergeLeaf() procedure, new nodes are not created but the edges of the existing ones are modified. The main advantage of this solution is a reduction in memory and an increase in functioning speed [17]; memory does not have to be allocated for a structure that represents a new node. In the worst case, a change in one edge near the terminal nodes causes the entire diagram to be recreated from the very beginning and memory to be allocated to all of its nodes. Modification of the existing edges has a certain disadvantage: in the MergeLeaf() procedure, ResultsTable cannot be employed since each pair of nodes $\mathbf{v_i}$, $\mathbf{v_j}$, may represent another extended equivalence class for each calling of the MergeLeaf($\mathbf{v_i}$, $\mathbf{v_j}$) procedure. In the Merge($\mathbf{v_i}$, $\mathbf{v_j}$) procedure, ResultsTable can be utilized since each node from the pair $\mathbf{v_i}$, $\mathbf{v_j}$ belongs to a path that leads to the same unambiguously determined node, which corresponds to a column pattern. For each column pattern, two additional nodes are added: **a**, **b**, **c**, **d**, **k**, **l**, **m**, and **n** processed by Merge(), and **a′**, **b′**, **c′**, **d′**, **k′**, **l′**, **m′**, and **n′** processed by MergeLeaf() (see Footnote 1) [2].

If the need to find equivalence classes exists for m functions $f_0, \ldots, f_{m\text{-}1}$, where $m > 2$, the two diagrams f'_0 and f'_1, are first combined and then the next diagram f'_2 is attached to the previously created diagram. In the worst case, the complexity of the presented algorithm does not exceed $O(n_0\, n_1 + n_1\, n_2 + \ldots + n_{\text{m-2}}\, n_{\text{m-1}})$, where n_{i} is the number of nodes (size) of the diagram f'_{i} (see Footnote 1) [2].

References

1. Opara A, Kubica M, Kania D (2019) Methods of improving time efficiency of decomposition dedicated at FPGA structures and using BDD in the process of cyber-physical synthesis. IEEE Access 7:20619–20631
2. Opara A, Kubica M, Kania D (2018) Strategy of logic synthesis using MTBDD dedicated to FPGA. Integr VLSI J 62:142–158
3. Kubica M, Opara A, Kania D (2017) Logic synthesis for FPGAs based on cutting of BDD. Microproces Microsyst 52:173–187
4. Opara A, Kubica M (2016) Decomposition synthesis strategy directed to FPGA with special MTBDD representation. In: International conference of computational methods in sciences and engineering. American Institute of Physics, Athens, 17 Mar 2016, Series: AIP conference proceedings, vol 1790
5. Lai Y, Pan KR, Pedram M (1996) OBDD-based function decomposition: algorithms and implementation. IEEE Trans Comput-Aided Des 15(8):977–990
6. Lai Y, Pan KR, Pedram M (1994) FPGA synthesis using function decomposition. In: Proceedings of the IEEE international conference on computer design, Cambridge, pp 30–35
7. Wurth B, Eckl E, Antreich K (1995) Functional multiple-output decomposition: theory and implicit algorithm. In: Design automation conference, pp 54–59

8. Scholl C, Molitor P (1994) Efficient ROBDD based computation of common decomposition functions of multi-output boolean functions. IFIP workshop on logic and architecture synthesis, Grenoble, pp 61–70
9. Scholl C (2001) Functional decomposition with application to FPGA synthesis. Kluwer Academic Publisher, Boston
10. Kubica M, Kania D (2015) SMTBDD : new concept of graph for function decomposition. IFAC conference on programmable devices and embedded systems PDeS, Cracow 2015, VII, 519, pp 61–66
11. Kubica M, Kania D (2016) SMTBDD: new form of BDD for logic synthesis. Int J Electron Telecommun 62(1):33–41
12. Kubica M, Kania D (2017) Area-oriented technology mapping for LUT-based logic blocks. Int J Appl Math Comput Sci 27(1):207–222
13. Kubica M, Kania D (2017) Decomposition of multi-output functions oriented to configurability of logic blocks. Bull Polish Acad Sci—Tech Sci 65(3):317–331
14. Kubica M, Kania D (2019) Technology mapping oriented to adaptive logic modules. Bull Polish Acad Sci Tech Sci 67(5):947–956
15. Kania D (2012) Programmable Logic Devices, PWN, Warszawa 2012 (in Polish)
16. Bryant RE (1986) Graph-based algorithms for Boolean function manipulation. IEEE Trans Comput C-35(8):677–691. Available: citeseer.ist.psu.edu/bryant86graphbased.htm
17. Kubica M, Kania D, Opara A (1790) Decomposition time effectiveness for various synthesis strategies dedicated to FPGA structures. In: International conference of computational methods in sciences and engineering. American Institute of Physics, Athens, 2016.03.17, Seria: AIP Conference Proceedings; vol 1790

Chapter 10
Ability of the Configuration of Configurable Logic Blocks

Configurable logic blocks (CLBs) are the main logic resources of an FPGA. In general, a CLB consists of a few logic cells. A typical elementary cell is based on LUTs. Currently, the functionality of configurable logic blocks, especially the number of inputs of LUTs, can be modified. In the XC3000 CLB [1], a single 5-input LUT (LUT5/1) or two 4-input LUTs (LUT4/2) with shared inputs are implemented. In Spartan [2], a similar configuration of the CLB is possible; however, the inputs of LUT4/2 are independent. In the most technologically advanced FPGAs, very flexible blocks, such as the ALM [3], are embedded.

10.1 Configuration Features of Logic Cells

Configuration capabilities of contemporary logic cells have been described in many scientific papers [4–8]. One of the characteristic features of these cells, which enables better configurability, is a considerably higher number of inputs compared with older constructions. Blocks that include a minimum of seven inputs are widely available [9]. The following example of ALM-based blocks included in the popular FPGA Stratix series by Altera shows the configuration abilities of modern CLBs. Blocks of this FPGA may be configured in six different ways, as illustrated in Fig. 10.1 [10].

The blocks may be configured in many ways. However, based on an analysis of configurabilities of ALM blocks by Intel, configurations may be divided into three groups. The first group involves configurations that have two independent LUTs with no common inputs. Appropriate LUTs include a stable number of inputs (appropriate values of k_a and k_b) and the implementation of separate (single) functions, as shown in Fig. 10.2a. The second group consists of configurations, in which some parts of the inputs must be shared among particular LUTs. Figure 10.2b illustrates the number of common inputs, which are denoted as k_{ab}. The third group includes configurations, in which the entire ALM block implements only one function, which usually has a

M. Kubica et al., *Technology Mapping for LUT-Based FPGA*, Lecture Notes in Electrical Engineering 713, https://doi.org/10.1007/978-3-030-60488-2_10

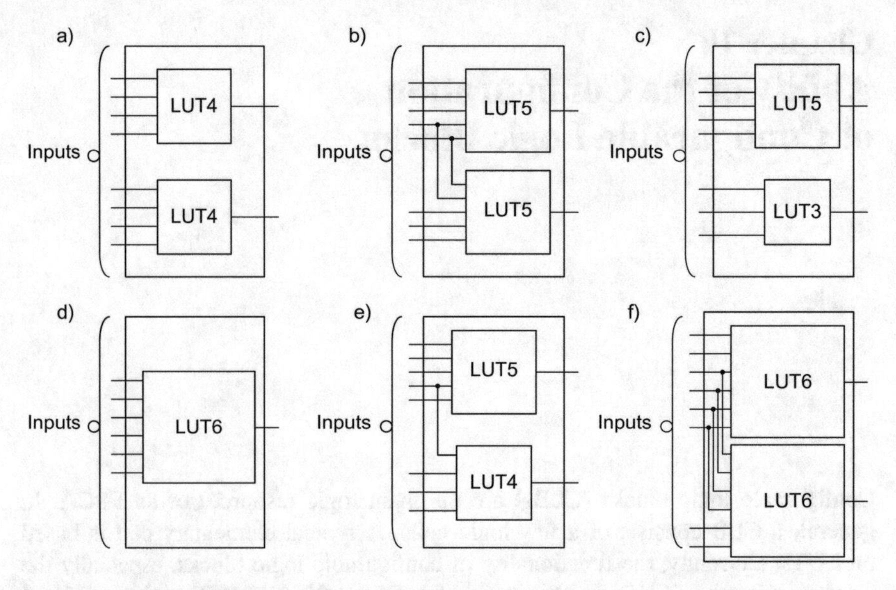

Fig. 10.1 ALM-based blocks configurations [3]

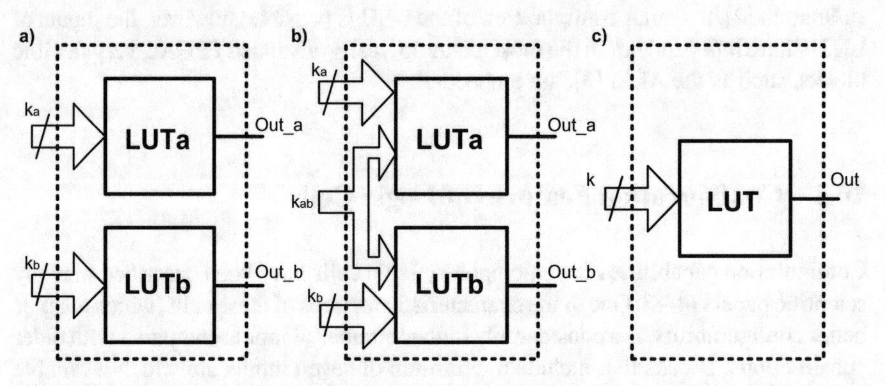

Fig. 10.2 Groups of configurations of logic blocks in an ALM: **a** a group without shared inputs; **b** a group with shared inputs; and **c** a group directed at implementing single functions

higher number of variables (k) than in previous cases. The third case is presented in Fig. 10.2c. [10]

The flexibility of logic blocks means that matching functions to ALM blocks that are directly connected with the decomposition process of logic functions is extremely important. Thus, choosing an appropriate configuration for an ALM block that is suitable for a possible partition of a circuit is crucial.

References

1. Xilinx, XC3000 Technical Information, xapp024, 1997
2. Xilinx Spartan-3 Generation FPGA User Guide (UG331), 2011
3. Altera (2012) Logic array blocks and adaptive logic modules in Stratix V Devices
4. Anderson J, Wang Q, Ravishankar C (2012) Raising fpga logic density through synthesis-inspired architecture. IEEE Trans Very Large Scale Integr (VLSI) Syst 20(3):537–550
5. Garg V, Chandrasekhar V, Sashikanth M, Kamakoti V (2005) A novel CLB architecture and circuit packing algorithm for logic-area reduction in SRAM-based FPGAs. Design automation conference, 2005. Proceedings of the ASP-DAC 2005. Asia and South Pacific, vol 2, pp 791–794
6. Manohararajah V, Singh DP, Brown SD (2005) Post-placement BDD-based decomposition for FPGAs. In: International conference on field programmable logic and applications, pp 31–38
7. Mao Z, Chen L, Wang Y, Lai J (2011) A new configurable logic block with 4/5-input configurable LUT and fast/slow-path carry chain. 2011 IEEE 9th International conference on ASIC (ASICON), pp 67–70
8. Rohani A, Zarandi H (2009) A new CLB architecture for tolerating SEU in SRAM-based FPGAs. ReConFig '09. In: International conference on reconfigurable computing and FPGAs, pp 83–88
9. Lattice (2012) Lattice ECP3 Family Data Sheet
10. Kubica M, Kania D (2017) Area-oriented technology mapping for LUT-based logic blocks. Int J Appl Math Comput Sci 27(1):207–222

Chapter 11
Technology Mapping of Logic Functions in LUT Blocks

The main goal of decomposition is an effective technology mapping of a function in an FPGA [1–12]. A multioutput function is mapped to logic blocks, which in the simplest case can be simply LUTs. These blocks can carry out any logic function with a limited (usually small) number of variables. The proposed technology mapping is based on choosing cutting lines of BDD.

11.1 Selection of Cutting Line

Let k be the number of logic block inputs. The essence of decomposition is the choice of an appropriate cutting line in BDDs. In the case of a simple serial decomposition, a cutting line should be chosen on the k-th level from a root, which is clearly depicted in Fig. 11.1. This choice of the cutting line makes the cardinality of bound set equal to k $(card(Xb) = k)$. In this way, all LUT inputs are utilized. The choice of decomposition is based on finding the cutting of a diagram that will guarantee a minimal number of bound functions. To find a solution with the lowest number of bound functions, various variable orderings in a BDD diagram should be analyzed[1] [13].

In the case of the multiple decomposition performed using the multiple cutting method, cutting levels should be chosen in such a way that the number of elements of separate bound sets ($Xb0$, …, Xbn) correspond to those of inputs of LUTs for the chosen configuration ($k0$,…, kn). The idea of this type of cutting is presented in Fig. 11.2 (see Footnote 1) [13].

The following example shows that the choice of the cutting level is essential from the point of view of the number of bound functions.

We consider the function described by the diagram illustrated in Fig. 11.3a. The FPGA contains LUTs with four inputs and one output (LUT4/1). The number of function variables is equal to 7. In this case, a diagram should be cut twice. Figure 11.3a

M. Kubica et al., *Technology Mapping for LUT-Based FPGA*, Lecture Notes in Electrical Engineering 713, https://doi.org/10.1007/978-3-030-60488-2_11

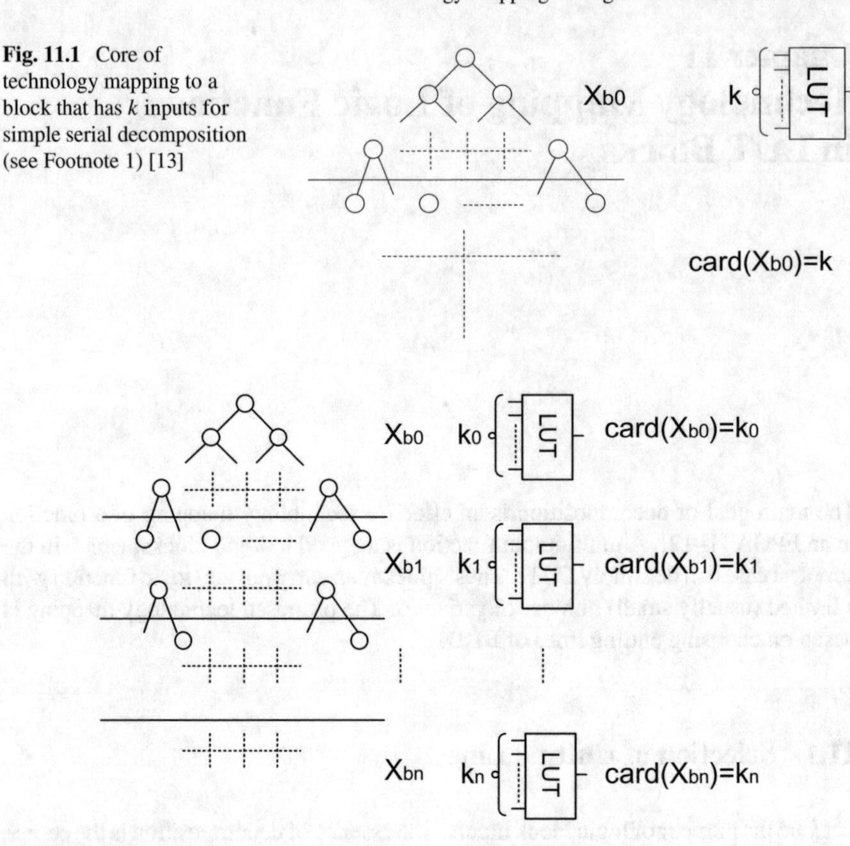

Fig. 11.1 Core of technology mapping to a block that has *k* inputs for simple serial decomposition (see Footnote 1) [13]

Fig. 11.2 Technology mapping for decomposition performed using several cutting lines (*X*b0,…, *Xbn*): separate bound sets (see Footnote 1) [13]

presents two alternative ways to cut a diagram. The first way (red and black cutting lines) causes variables' partitioning into the sets Xb0 = $\{x_0, x_1, x_2\}$ and Xb1 = $\{x_3, x_4, x_5, x_6\}$. The second cutting (blue and black lines) divides variables into the sets Xb0 = $\{x_0, x_1, x_2, x_3\}$ and Xb1 = $\{x_4, x_5, x_6\}$ (see Footnote 1) [14].

In the case of searching for effective multiple decomposition, the number of bound functions needs to be determined for all parts of subdiagrams. In the Xb0 part, the number of cut nodes must be determined. Regarding the Xb1 parts, the column multiplicity of the root tables illustrated in Fig. 11.3d, e must be determined. If the number of bound functions for two cuttings is known, the process of creating a circuit shown in Fig. 11.3b, c is possible. A better solution is obtained after the first cutting (Fig. 11.3b). The bound block from Fig. 11.3b requires the use of three LUT4/1; in the case of Fig. 11.3c, 5 blocks are required. Cutting lines, such as card(Xb0) = 3 and card(Xb1) = 4 provide a better solution with respect to the number of LUTs (see Footnote 1) [14].

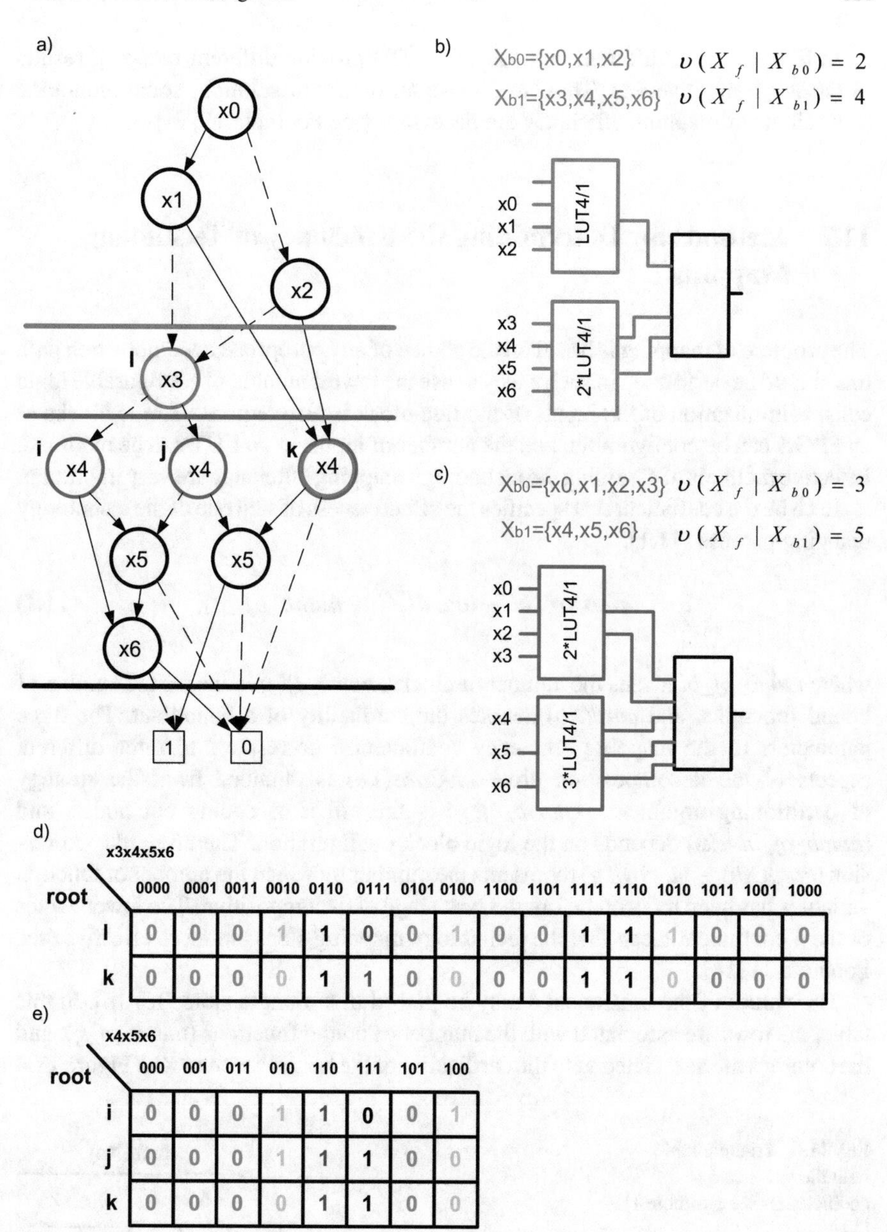

Fig. 11.3 Multiple decomposition performed using the method of multiple cutting; **a** ROBDD diagram subjected to multiple cuttings, **b** blocks (red) related to the first logic level for a red cutting line, **c** blocks (blue) related to the first logic level for a blue cutting line, **d** root table associated with SMTBDD between a black cutting line and a red cutting line, and e) root table associated with SMTBDD between a black cutting line and a blue cutting line (see Footnote 1) [14]

In this example, different cuttings of a BDD provide different mapping results in terms of effectiveness. Thus, to choose an optimum solution, some monotone coefficients of mapping efficiency are necessary (see Footnote 1) [14].

11.2 Methods for Determining the Efficiency of Technology Mapping

The problem of mapping is based on the choice of an appropriate decomposition path that should be performed in such a way to use the lowest number of configurable logic cells. Minimization of the inputs of the free block is also required. Logic blocks of an FPGA can be configurable, i.e., the number of inputs to an LUT can be modified. Effective methods of assessing the technology mapping efficiency are very important.

Let δ be the coefficient that specifies the effectiveness of step one of the technology mapping process (11.1),

$$\delta = numb\_of\_bl - (card(Xb) - numb\_of\_g), \tag{11.1}$$

where *numb_of_bl* means the number of blocks, *numb_of_g* indicates the number of bound functions, and *card(Xb)* denotes the cardinality of a bound set. The three parameters of the mapping efficiency coefficient δ correspond to three different aspects of the decomposition process. *Card*(*Xb*) is obtained from the strategy of partitioning arguments, (*numb_of_g*) is the effect of coding cut nodes, and (*numb_of_blocks*) depends on the logic block configuration. Therefore, the expression (*card*(*Xb*) – *numb of g*) represents the number by which the number of function variables has been reduced before the next stage of decomposition. The lower values of the δ coefficient mean that the technology mapping steps are more effective (see Footnote 1) [14].

The values of the coefficient δ may be placed in a triangle table [15, 16]. In this table, the rows are associated with the number of bound functions (*numb_of_g*), and the columns are associated with the cardinality of the bound set *card(Xb)*. Figure 11.4

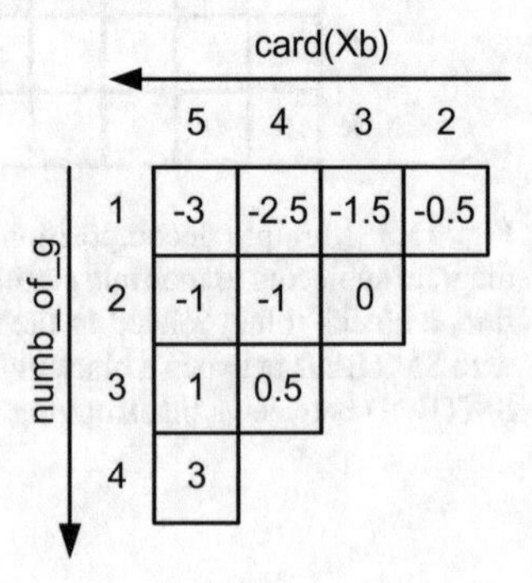

card(Xb)
5
4
3
2
numb_of_g
1
2
3
4
-3
-2.5
-1.5
-0.5
-1
-1
0
1
0.5
3

Fig. 11.4 Triangle table with the value of δ coefficients (see Footnote 1) [14]

presents an example of a triangle table with δ coefficients for logic blocks that may be configured as the LUT5/1 or LUT4/2 (LUTn/m – n inputs and m outputs LUT) (see Footnote 1) [14].

In this method, the number of bound functions (*numb_of_g*) is the parameter that determines the 'quality' of a variable ordering. The number of elements of a bound set (*card(Xb)* = *k* is assumed) is associated with possible configurations of the number of inputs in a logic block. The values of an efficiency cofactor of the mapping δ are placed in the cells of a table and take into account the third parameter - the number of blocks (*numb_of_bl*) needed to carry out a given decomposition. Assessing the efficiency of mapping on a given decomposition stage is based on determining the value of the δ cofactor based on the pair of parameters (*card(Xb), numb_of_g*). The value of the coefficient δ enables choosing the optimal level of the cutting of a BDD diagram.

We consider technology mappings of the function described by the BDD diagram. To find decomposition and guarantee the best technology mapping to the blocks LUT5/1 or LUT4/2, three cutting lines should be analyzed on the following levels: 3 (Fig. 11.5a), 4 (Fig. 11.5b), and 5 (Fig. 11.5c), including the root; bound sets contain 3, 4, and 5 variables, respectively. A triangle table, in which the symbol "O" is indicated for the given *card(Xb)* and *numb_of_g* of the obtained value δ, is associated with each BDD (see Footnote 1) [14].

Searching for decomposition should be started with the analysis of a variable partitioning that corresponds to the lowest value δ = −3. This value indicates decompositions associated with an ordered pair of numbers (*card(Xb), numb_of_g*) = (5,1). The value δ occurs only in the case of a five-element bound set (Fig. 11.5c). Thus, searching should be started with the *card(Xb)* = 5. For the diagram in Fig. 11.5c, five cut nodes exist. To distinguish these nodes, the use of three bound functions is necessary. The value of the coefficient δ for decomposition (*card(Xb), numb_of_g*) = (5,3) is 1; thus, it is higher from the minimal value equal to - 3. This solution is marked in a triangle table (Fig. 11.5c) with the symbol 'O' (see Footnote 1) [14].

Variables' partitions, for which *card(Xb)* < 5 are analyzed. For *card(Xb)* < 5, the minimal value δ = −2.5 is obtained in the case (*card(Xb), numb_of_g*) = (4, 1). Unfortunately, the diagram in Fig. 11.5b has three cut nodes. Therefore, two bound functions are needed. The value of δ = -1 is obtained for (*card(Xb), numb_of_g*) = (4, 2) is lower than the value of the coefficient in the previous step of the analysis. The analyzed variables' partitioning is more effective when the mapping efficiency is taken into account. The obtained solution is marked with the symbol 'O' in a triangle table in Fig. 11.5b.

A three-element bound set is analyzed. Only one case in which δ < -1 occurs for decomposition (*card(Xb), numb_of_g*) = (3, 1). The cutting of a diagram (Fig. 11.5a) requires two bound functions; the solution does not exceed the solution obtained in the previous steps (see Footnote 1) [14].

In this example, decomposition, for which (*card(Xb), numb_of_g*) = (4, 2), guarantees the best mapping.

The proposed technique can be expanded to other logic blocks and for the multiple cutting method using a SMTBDD diagram [13, 17, 18].

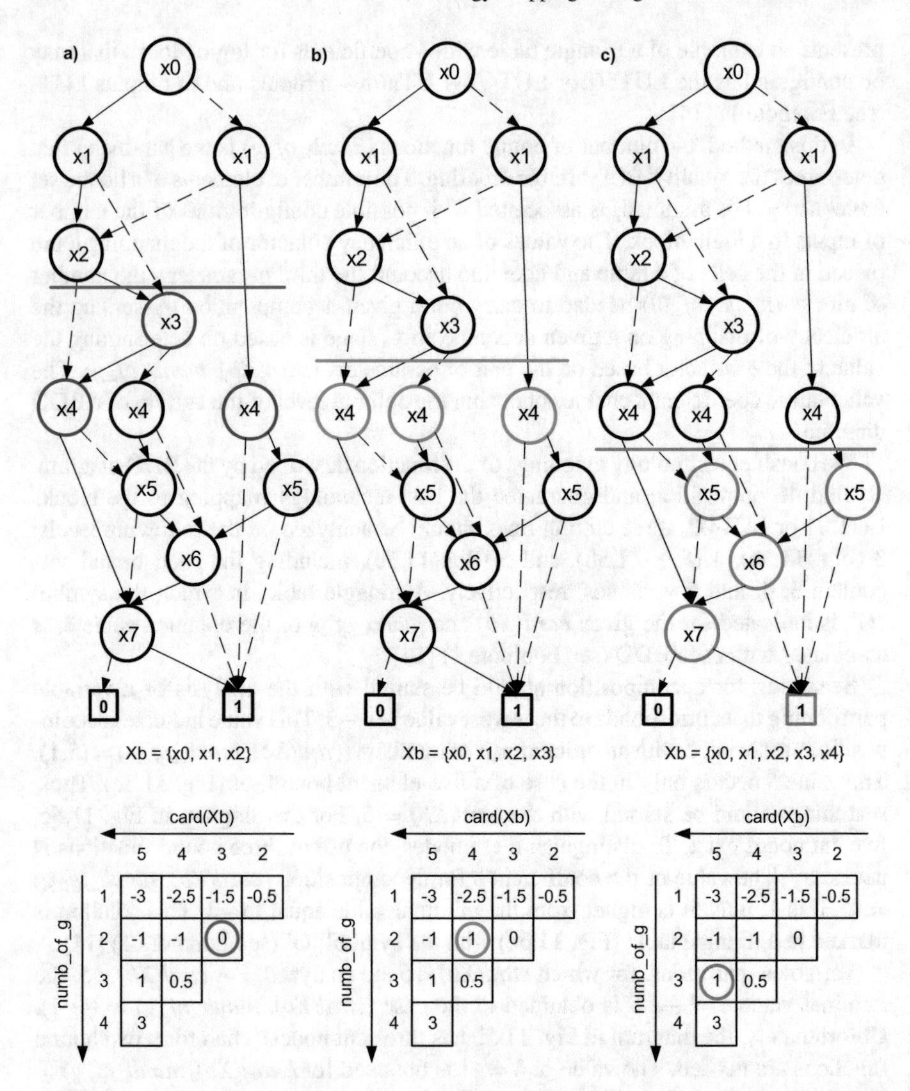

Fig. 11.5 One step of decomposition of a logic function with triangle tables: **a** the cutting line on level 3, **b** the cutting line on level 4, and **c** the cutting line on level 5 (see Footnote 1) [14]

Many methods can be applied to determine triangular tables [14, 16]. Figure 11.6 shows 4 examples of triangular tables; the advantages and disadvantages are discussed in the next section.

The value of the cofactor δ in the table in Fig. 11.6a is calculated as the difference between the number of bound functions and the cardinality of a bound set. The solution associated with the minimum value δ is sought and corresponds to the best technology mapping. This case is the most basic case and enables a mapping of the function using LUTs with five inputs and one output. In this case, one LUT5/1 block

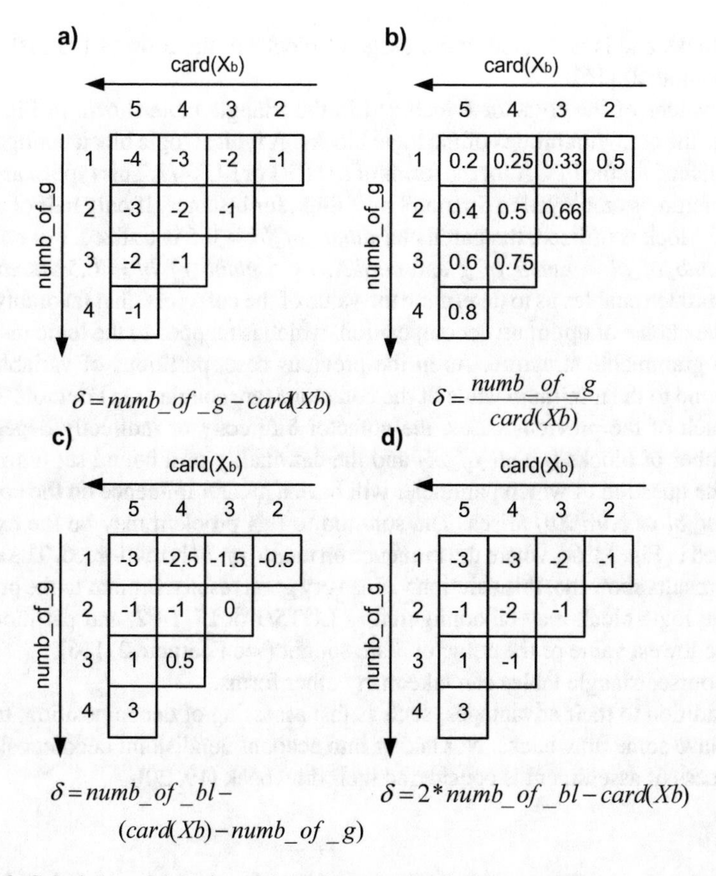

Fig. 11.6 Calculation of the value of the cofactor δ: triangle tables (see Footnote 2) [16]

is necessary to carry out each available decomposition. For each decomposition [in addition to the case described using the ordered pair (5,1)], a part of the input of the LUT is not employed, indicating that the resources of the LUT are irrecoverably lost. The main disadvantage of the efficient technology mapping that is performed using a triangle table, as shown in Fig. 11.6a, is that the value of δ is the same for many partitions of variables. This condition prevents us from choosing the best decomposition and may yield solutions that are not optimal [2] [16].

The method of determining the effectiveness of technology mapping based on the formula presented below the triangle table in Fig. 11.6b does not have this drawback. In this case, the value of the cofactor δ is determined based on the quotient of two parameters: *numb_of_g* and *card*(Xb). Similar to the previous case, this method searches for partitions for which the value of the cofactor δ takes a minimum value. However, this approach also fails to take into account the reconfigurabilities of the logic blocks and is dedicated to the usage of blocks in the form of LUT5/1 blocks (see Footnote 2) [16].

The values of the cofactor δ included in the triangle table shown in Fig. 11.6c consider the configurabilities of the logic blocks. A typical logic block configuration is established for the FPGA for the forms of LUT5/1 or LUT4/2. This type of architecture/structure is employed in Spartan3 by Xilinx, for instance. If only half of a given LUT4/2 block is utilized, the parameter *numb_of_bl* = 0.5 is utilized. For *card(Xb)* = 5: *numb_of_bl* = *numb_of_g* and *card(Xb)* < 5 *numb_of_bl* = 0.5**numb_of_g*. This approach enables us to determine the value of the cofactor δ that favorably influences the choice of optimum decomposition, which is mapped to the logic resources of a programmable structure. As in the previous case, partitions of variables that correspond to the minimum value of the cofactor δ, are sought (see Footnote 2) [16].

In each of the previous cases, the cofactor δ directly or indirectly depends on the number of blocks (*numb_of_bl*) and the cardinality of a bound set (*card(Xb)*). Thus, the question of which parameter will have a greater influence on the cofactor: *numb_of_bl* or *card(Xb)* arises. The solution to this problem may be the example presented in Fig. 11.6d, where the influence on *numb_of_bl* is reinforced. The experimental results show that this solution yields very good results. Similar to the previous case, the logic block may be configured as LUT5/1 or LUT4/2, and partitions that have the lowest value of the cofactor δ are sought (see Footnote 2) [16].

Of course, triangle tables can take many other forms.

In addition to their advantages, such as fast assessing of decomposition, triangle tables have some drawbacks. Not taking into account nondisjoint decomposition in the process of assessment is considered their drawback [19, 20].

11.3 Triangle Tables, Including Nondisjoint Decomposition

To introduce the nondisjoint decomposition, the triangle tables shown in the papers must be modified [21–23]. The modification consists in adding a third dimension describing card(Xs). Figure 11.7 shows a set of tables associated with various values of the number of elements of a shared set Xs [23].

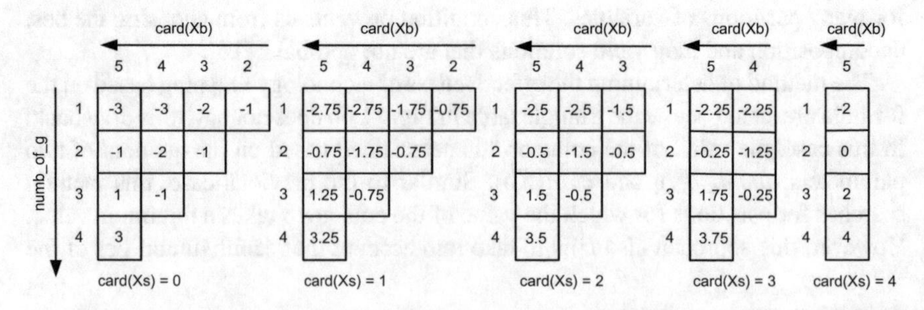

Fig. 11.7 Set of tables that describe a three-dimensional triangle table that takes into account the number of the set *Xs* (© (2020) IEEE. Reprinted with permission from Opara et al. [23])

Figure 11.7 shows a three-dimensional triangle table whose cells contain value of the cofactor δ. In general, the lower is the value of the cofactor δ, the better is the technology mapping. The value of the cofactor δ may be presented using the following expression (11.2):

$$\delta = 2 * numb\_of\_bl - card(Xb) + card(Xs)/4 \tag{11.2}$$

It is noteworthy that the parameter numb_of_g does not appear in the above equation. In its place the parameter numb_of_bl was introduced, which for LUT5 is: numb_of_g = numb_of_bl and in the case of 2*LUT4: numb_of_g = 2 * numb_of_bl. These configurations are associated with devices Spartan 3 [24]. The use of three-dimensional triangle tables to assess a technology mapping is illustrated in Fig. 11.8 [23].

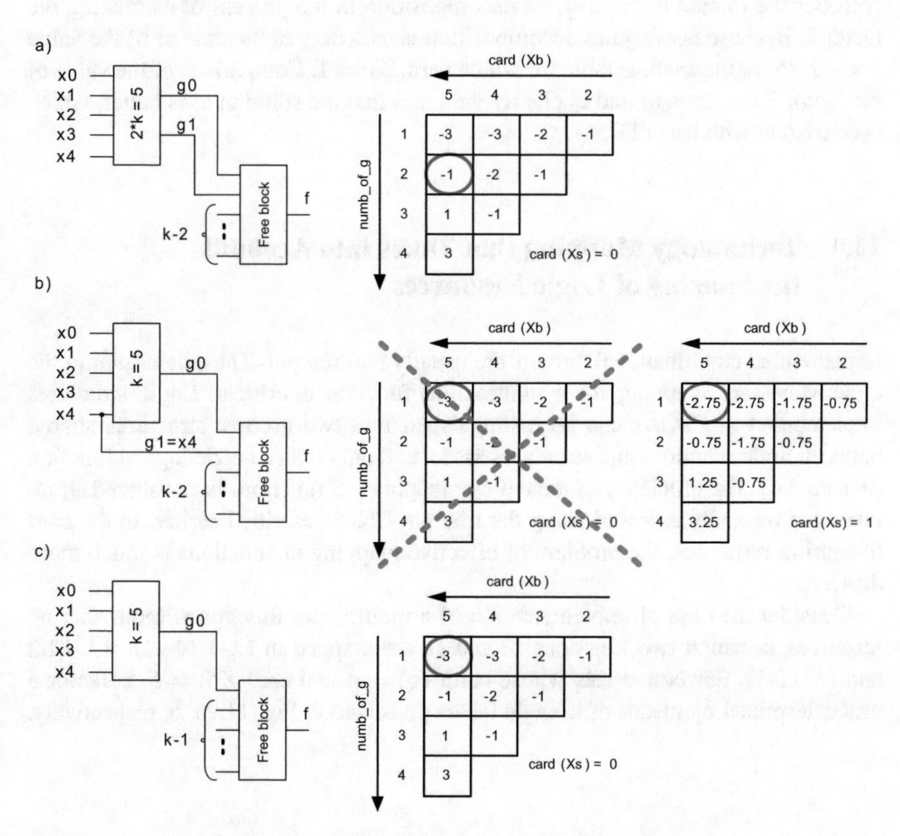

Fig. 11.8 Using three dimensional triangle tables to assess a technology mapping (© (2020) IEEE. Reprinted with permission from Opara et al. [23])

Consider the following case in which three technology mappings are presented in the form of triangle tables as shown in Fig. 11.8. First, consider the situation in which we do not include in the process of determining δ the occurrence of the nondisjoint decomposition card(Xs) = 0. In the case of the triangle table shown in Fig. 11.8a, two bund functions are necessary and the value $\delta = -1$. In the case of the triangle table shown in Fig. 11.8b, the value $\delta = -3$ (when estimating δ the influence of nondisjoint decomposition was not taken into account). In the case of Fig. 11.8c, also the value $\delta = -3$, however, in this case there is no nondisjoint decomposition. There is a problem which of the solutions b) or c) is better. In both cases the value is $\delta = -3$, therefore the presented approach does not solve this problem. Now consider the number of free block inputs. It turns out that in the case of b) the free block has $k - 2$ unused inputs, while in the case of c) the number of unused free block inputs is $k - 1$, which means that the solution c) is a more effective solution. Let us now consider the impact of nondisjoint decomposition in the process of estimating the factor δ. Because nondisjoint decomposition occurs only in the case of b) the value $\delta = -2.75$ for the triangle table for which card(Xs) = 1. Comparison of the value of the factor δ for cases b) and c) clearly indicates that the solution c) is better, which is consistent with true [23].

11.4 Technology Mapping that Takes into Account the Sharing of Logic Resources

Implemented combinational circuits are usually multioutput. Thus, determining the effectiveness of mapping for a multioutput function is critical. Logic structures implemented in FPGAs can be distinguished into two groups: structures shared between logic functions and structures associated only with a single logical function (nonshared). The efficiency of technology mapping of functions in nonshared structures can be easily assessed using the triangle tables described earlier. In the case of sharing resources, the problem of effective mapping of functions is much more difficult.

Consider the case of implementation of a multioutput function without sharing resources in which two functions f0 and f1 are mapped in LUT blocks (LUT4/2 and LUT5/1). For bound sets whose card(Xb) = 4 and card(Xb) = 5, cofactor δ was determined by means of triangle tables presented in Fig. 11.9a, b, respectively.

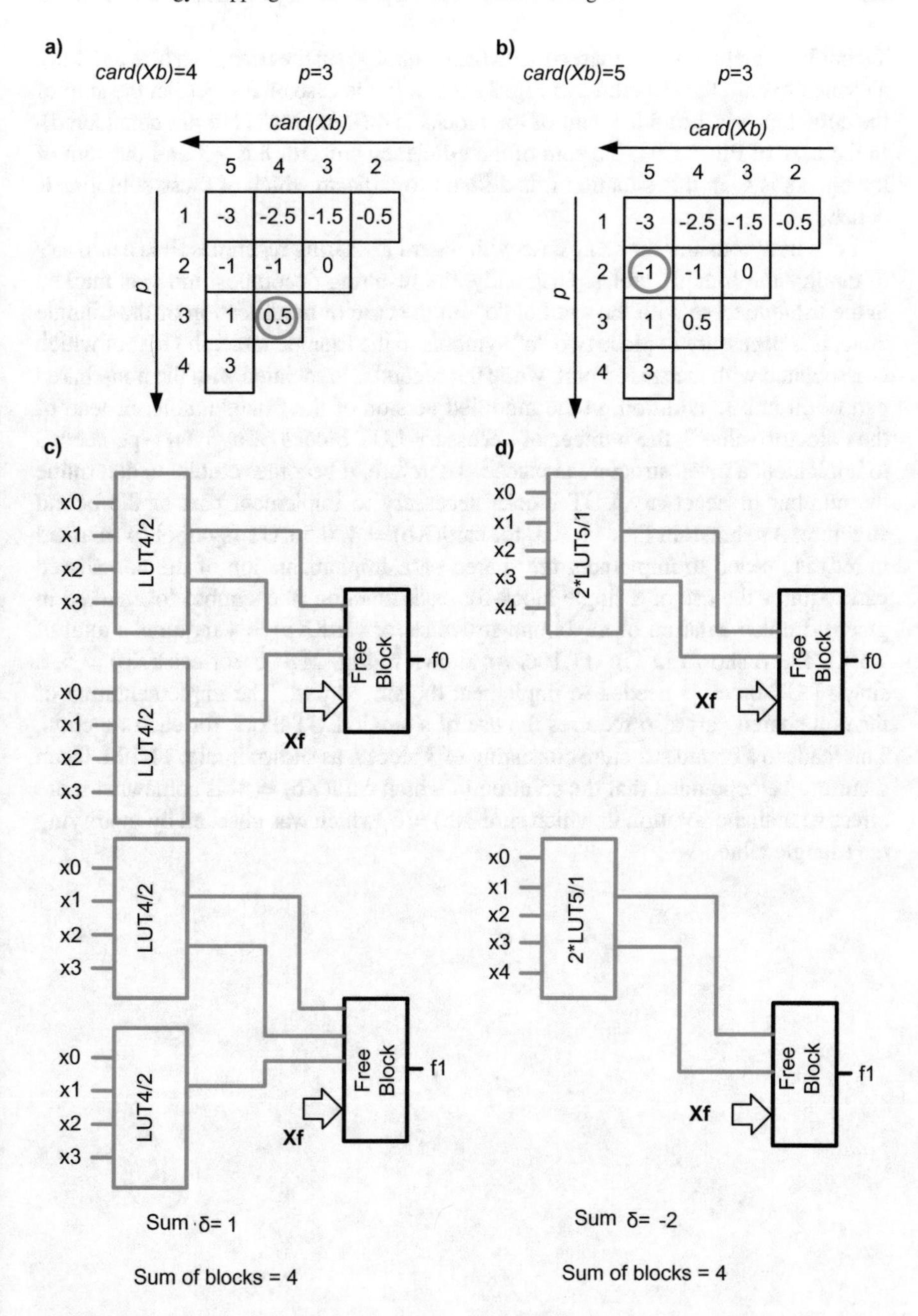

Fig. 11.9 Triangle tables and gained structures (without considering blocks' sharing); **a** a triangle table for card(Xb) = 4, **b** a triangle table for card(Xb) = 5, and **c**, **d** gained structures

Technology mappings are marked with the symbol ‘o’ in the triangle tables and have a chance to gain the structures in Fig. 11.9c, d. In the case of Fig. 11.9c, the sum of the cofactors δ is 1 and the sum of the blocks is 4 (free blocks are not considered). In the case of Fig. 11.9d, the sum of the efficiency cofactor δ is -2 and the sum of the blocks is 4. In this situation it is difficult to indicate which of these solutions is better.

Let’s now consider the same case with regard to sharing resources. It is necessary to modify the triangle tables. Originally, the resulting decomposition was marked in the triangle table with the symbol “o". In the case of modification of the triangle table, it is necessary to place two “o” symbols in the triangle table, the first of which is associated with the shared part, while the second is associated with the non-shared part of circuit. In addition, in the modified version of the triangle table, instead of the cofactor value δ, the number of necessary LUT blocks of a given type needed to implement a given structure is placed. Therefore, it becomes crucial to determine the number of necessary LUT blocks necessary to implement part of the bound structure. As shown in Fig. 11.10a, for card(Xb) = 4, 0.5 LUT (symbol ‘o’ marked in red) is needed to implement the shared part. Implementation of the non-shared part requires the use of a single block for each function (the symbol ‘o’ marked in green). Implementation of the bound structure for card(Xb) = 4 requires a total of 2.5 LUTs, as shown in Fig. 11.10c. As shown in Fig. 11.10b, for card(Xb) = 5, a single LUT block is needed to implement the shared part. The implementation of the non-shared part also requires the use of a single LUT block for each function. This leads to a bound structure consisting of 3 blocks, as shown in Fig. 11.10d. It can therefore be concluded that the solution in which card(Xb) = 4, is somewhat more effective than the solution in which card(Xb) = 5, which was checked by modifying the triangle table.

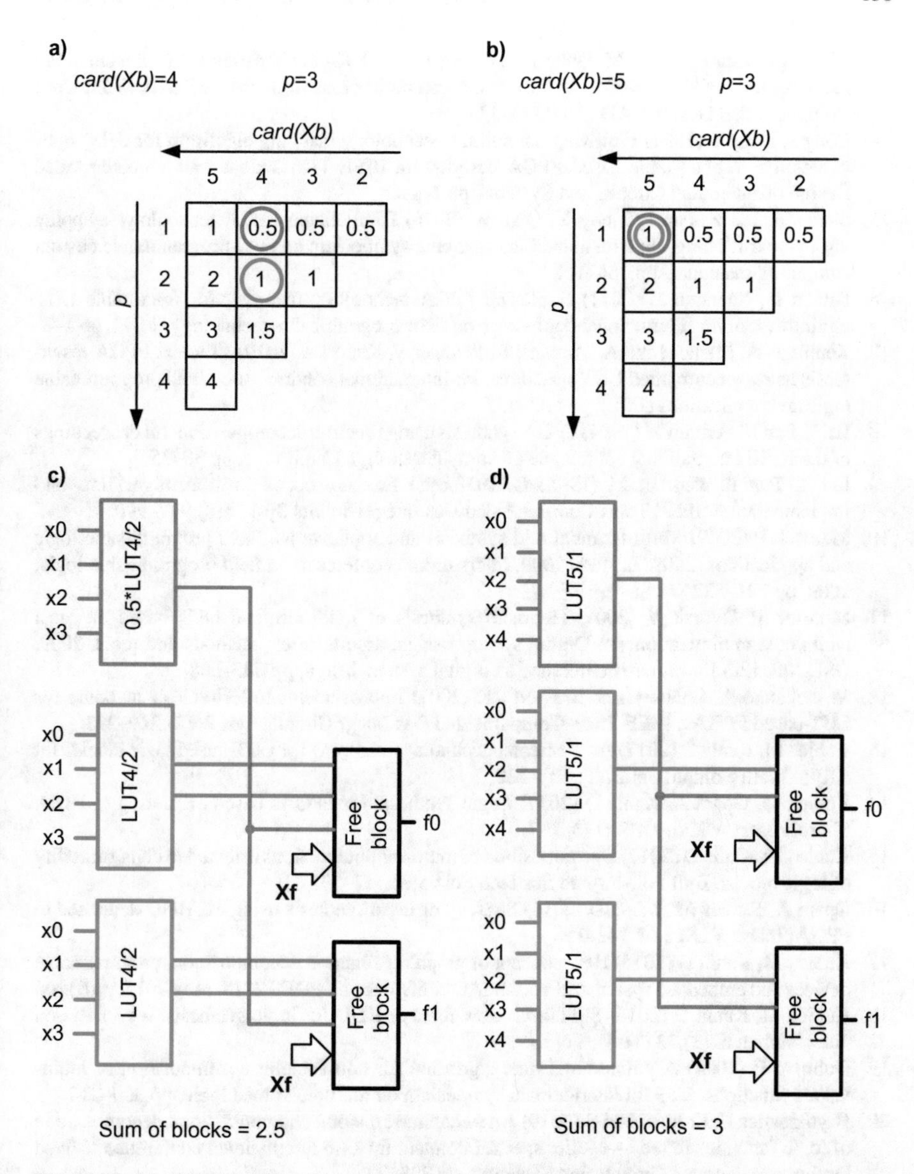

Fig. 11.10 Modified triangle tables and obtained mappings (sharing blocks is taken into account)

References

1. Chang S, Marek-Sadowska M, Hwang T (1996) Technology mapping for TLU FPGA's based on decomposition of binary decision diagrams. IEEE Trans Comput-Aided Des 15(10):1226–1235
2. Chávez RS, Rajavel ST, Akoglu A (2012) WL-Emap: Wirelength prediction based technology mapping for FPGAs. In: VIII Southern Conference on Programmable Logic (SPL), pp 1–6

3. Cheng L, Chen D, Wong M, Hutton M, Govig J (2007) Timing constraint-driven technology mapping for FPGAs considering false paths and multi-clock domains. IEEE/ACM Int Conf Comput-Aided Des, ICCAD 2007:370–375
4. Cong J, Ding Y (1994) FlowMap: an optimal technology mapping algorithm for delay optimization in lookup-table based FPGA designs. In: IEEE Transactions on Computer-Aided Design of Integrated Circuits and Systems, pp 1–12
5. Shen Ch, Lin Z, Fan P, Meng X, Qian W (2016) Parallelizing FPGA technology mapping through partitioning. In: 24th annual international symposium on field-programmable custom computing machines, pp 164–167
6. Dickin D, Shannon L (2011) Exploring FPGA technology mapping for fracturable LUT minimization. In: International conference on field-programmable technology (FPT), pp 1–8
7. Kennings A, Mishchenko A, Vorwerk K, Pevzner V, Kundu A (2010) Efficient FPGA resynthesis using precomputed LUT structures. In: International conference on field programmable logic and applications (FPL), pp 532–537
8. Lai Y, Pan K, Pedram M (1994) FPGA synthesis using function decomposition. In: Proceedings of the IEEE international conference on computer design, Cambridge, pp 30–35
9. Lai Y, Pan K, Pedram M (1996) OBDD-based function decomposition: algorithms and implementation. IEEE Trans Comput Aided Des Integr Circuits Syst 15(8):977–990
10. Mikusek P (2009) Multi-terminal bdd synthesis and applications, Field programmable logic and applications, 2009. In: FPL 2009. International conference on field programmable logic, 2009, pp 721–722
11. Mikusek P, Dvorak V (2009) Heuristic synthesis of multi-terminal bdds based on local width/cost minimization. In: Digital system design, architectures, methods and tools, 2009. DSD '09. 12th Euromicro conference on digital system design, pp 605–608
12. Mishchenko A, Chatterjee S, Brayton RK (2007) Improvements to technology mapping for LUT-based FPGAs. IEEE Trans Comput Aided Des Integr Circuits Syst 26(2):240–253
13. Kubica M, Kania D (2017) Area-oriented technology mapping for LUT-based logic blocks. Int J Appl Math Comput Sci 27(1):207–222
14. Kubica M, Opara A, Kania D (2017) Logic synthesis for FPGAs based on cutting of BDD. Microprocess Microsyst 52:173–187
15. Kubica M, Kania D (2017) Decomposition of multi-output functions oriented to configurability of logic blocks. Bull Polish Acad Sci Tech Sci 65(3):317–331
16. Opara A, Kubica M, Kania D (2018) Strategy of logic synthesis using MTBDD dedicated to FPGA. Integr VLSI J 62:142–158
17. Kubica M, Kania D (2015) New concept of graph for function decomposition, programmable devices and embedded systems, 2015. In: IAC conference on PDES 2015, May 2015, pp 61–66
18. Kubica M, Kania D (2016) SMTBDD: new form of BDD for logic synthesis. Int J Electron Telecommun 62(1):33–41
19. Dubrova E (2004) A polinominal time algorithm for non-Disjoint decomposition of multi-valued functions. In: 34th international symposium on multiple-valued logic, pp 309–314
20. Hrynkiewicz E, Kołodziński S (2010) An Ashenhurst disjoint and non-disjoint decomposition of logic functions in reed—Muller spectral domain. In: 17th international conference "Mixed Design of Integrated Circuits and Systems", pp 293–296
21. Opara A, Kubica M (2017) Optimization of synthesis process directed at FPGA circuits with the usage of non-disjoint decomposition. In: Proceedings of the international conference of computational methods in sciences and engineering 2017. American Institute of Physics, Thessaloniki, 2017.04.21, Seria: AIP conference proceedings; vol 1906, Art. no. 120004
22. Opara A, Kubica M (2018) The choice of decomposition taking non-disjoint decomposition into account. In: Proceedings of the international conference of computational methods in sciences and engineering 2018. American Institute of Physics, Thessaloniki, 2018.03.14, Seria: AIP conference proceedings; vol 2040, Art. no. 080010
23. Opara A, Kubica M, Kania D (2019) Methods of improving time efficiency of decomposition dedicated at FPGA structures and using BDD in the process of cyber-physical synthesis. IEEE Access 7:20619–20631. https://doi.org/10.1109/ACCESS.2019.2898230
24. Xilinx (2011) Spartan-3 Generation FPGA, User Guide (UG331)

Chapter 12
Technology Mapping of Logic Functions in Complex Logic Blocks

In modern FPGA devices, the basic logic cell does not contain LUTs but does contain configuration logic blocks. These blocks contain several LUTs in their structure that are significantly linked to each other. The architecture of configurable logic blocks is focused on increasing the flexibility of these cells in terms of configuration possibilities. Therefore, the inclusion of these possibilities in the process of technology mapping is crucial [1–7]. This problem will be discussed using the example of Intel ALM blocks [8].

12.1 Technology Mapping in ALM Blocks

We consider two multioutput functions that are described using the MTBDD diagram in Fig. 12.1. To guarantee the best technology mapping to ALM blocks, the process of decomposition should be matched to possible configurations of ALM blocks and the configuration that can guarantee the most effective technology mapping of the analyzed multioutput function should be selected [9].

From the analyzed methods of multiple cutting from the BDD point of view, the configurations, in which the inputs of LUTs are not shared, are the most efficient. Searching for the effective creation of a bound block shall be performed using the model of multiple decomposition [10]. LUTs included in the ALM block may be associated with appropriate extracts of a diagram. An ALM block configuration that enables the best mapping of the results of multiple decomposition to the logic resources of a circuit is shown in Fig. 12.2 [9].

The number of inputs of separate LUT cells included in ALM blocks determine the cutting levels of a diagram, as shown in Fig. 12.1. The key task is to determine whether decomposition that can match a circuit into a single ALM block is possible. Therefore, the cases for the cuttings associated with the configuration presented in Fig. 12.2a is necessary (on levels 4 and 8 counting from the root), Fig. 12.2b (on levels 5 and 8 or 3 and 8 counting from the root), and Fig. 12.2c (on level 6 counting

M. Kubica et al., *Technology Mapping for LUT-Based FPGA*, Lecture Notes in Electrical Engineering 713, https://doi.org/10.1007/978-3-030-60488-2_12

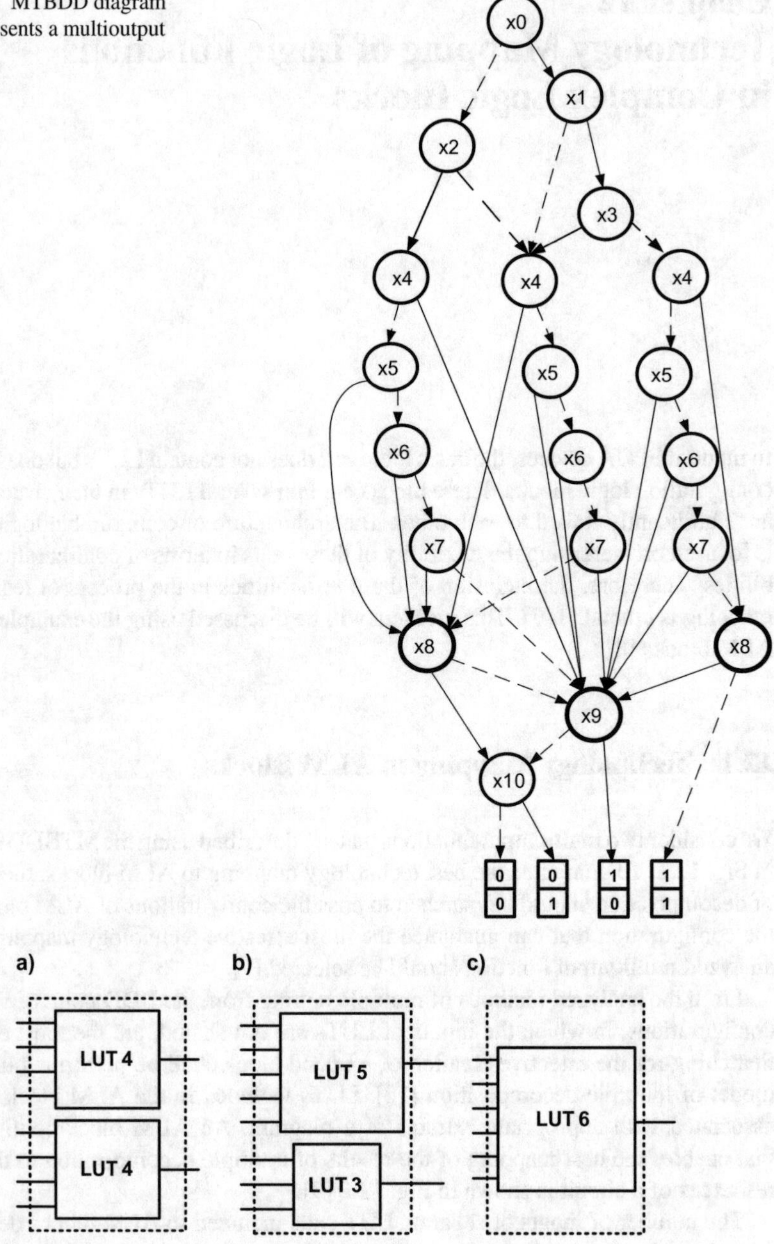

Fig. 12.1 MTBDD diagram that represents a multioutput function

Fig. 12.2 ALM block configurations that guarantee the best technology mapping in the case of multiple decomposition

from the root). Note that the configuration of an ALM block, as shown in Fig. 12.2c, enables only one extract of a diagram to be carried out [9].

The cutting of a diagram enables separate extracts in an ALM block to be configured as shown in Fig. 12.2a. Thus, the cutting of the MTBDD diagram on levels 4 and 8 counting from the root is analyzed and presented in Fig. 12.3a. For a zero extract, for which the set $X_{b0} = \{x_0, x_1, x_2, x_3\}$, two bound functions are necessary to distinguish the three cut nodes marked in blue. As the result of searching for nondisjoint decomposition, the variable x_0 may fulfill the role of a switching function. In this case, one bound function will be performed in the LUT4/1 block, and the second bound function will be replaced with the variable x_0. Searching for technology mapping of the first extract, for which $X_{b1} = \{x_4, x_5, x_6, x_7\}$, requires the analysis of a root table, as shown in Fig. 12.3b [9].

Column multiplicity of a root table is 2. Thus, only a single bound function is needed. A bound block connected with extract 1 may be performed in one LUT4/1 block, which is included in the ALM block.

As the result of a double cutting of the MTBDD diagram, the structure presented in Fig. 12.3c is created and performed in a single ALM block [9].

ALM blocks can be configured using other ways. Thus, other possible technology mappings should be considered. While analyzing the configuration shown in Fig. 12.2b, two other ways of MTBDD diagram cutting shall be considered. In the first way, the cutting lines are led on levels 3 and 8 counting from the root (Fig. 12.4a). In the second way, the cutting lines are led on levels 5 and 8 (Fig. 12.5a) [9].

In the first case, with the cutting lines, as presented in Fig. 12.4a, a zero extract ($X_{b0} = \{x_0, x_1, x_2\}$) has only one root. Thus, the number of bound functions depends on the number of cut nodes. The differentiation of three cut nodes enables technology mapping in which two bound functions exist. As in the previous case, the variable x_0 may fulfill the role of a bound function. Therefore, to carry out a bound block associated with this extract, the LUT3/1 block is sufficient and is the element of a complex configuration of an ALM block (Fig. 12.2b). For the first extract ($X_{b1} = \{x_3, x_4, x_5, x_6, x_7\}$), column multiplicity should be determined based on the root table shown in Fig. 12.4b. The column multiplicity of this table is 3; thus, two bound functions require two LUT5/1 blocks. In the ALM cell for the configuration shown in Fig. 12.2b, only one LUT block exists. Therefore, for a complex cutting of the MTBDD diagram, bound blocks cannot be carried out in a single ALM block. In this case, two ALM blocks are required, which is worse than the solution matched to the configuration shown in Fig. 12.2a [9].

An alternative method for cutting the MTBDD diagram is presented in Fig. 12.5a. The extract, which is associated with the set $X_{b0} = \{x_0, x_1, x_2, x_3, x_4\}$, is identified with the edges that originate from the nodes included in it and indicate six cut nodes. Creating a bound block in the form of a structure with three outputs (three bound functions) is necessary. For the cutting lines, no variable included in the extract 0 can fulfill the role of a switching function that enables nondisjoint decomposition. Thus, three LUT5/1 blocks are required. Extract 1 ($X_{b1} = \{x_5, x_6, x_7\}$) is connected to a root table, as shown in Fig. 12.5b, whose column multiplicity is 2. To carry out a bound block associated with it, the LUT3/1 block is sufficient. The obtained solution

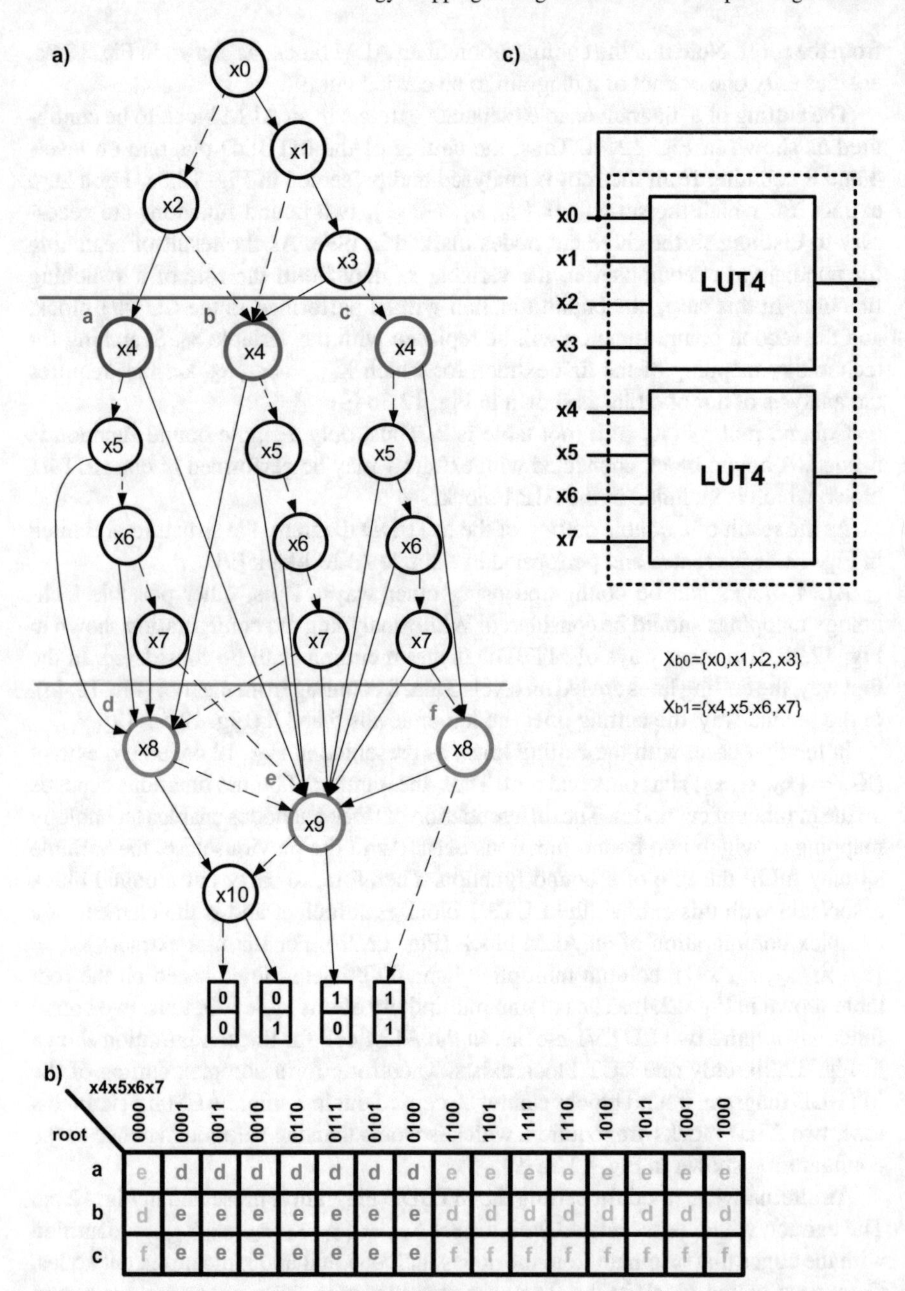

root	0000	0001	0011	0010	0110	0111	0101	0100	1100	1101	1111	1110	1010	1011	1001	1000
a	e	d	d	d	d	d	d	d	e	e	e	e	e	e	e	e
b	d	e	e	e	e	e	e	e	d	d	d	d	d	d	d	d
c	f	e	e	e	e	e	e	e	f	f	f	f	f	f	f	f

Fig. 12.3 Decomposition performed by multiple cuttings; **a** MTBDD diagram subjected to multiple cuttings, **b** a root table associated with the extract between cutting lines, **c** the obtained structure that corresponds to bound blocks inside an ALM block

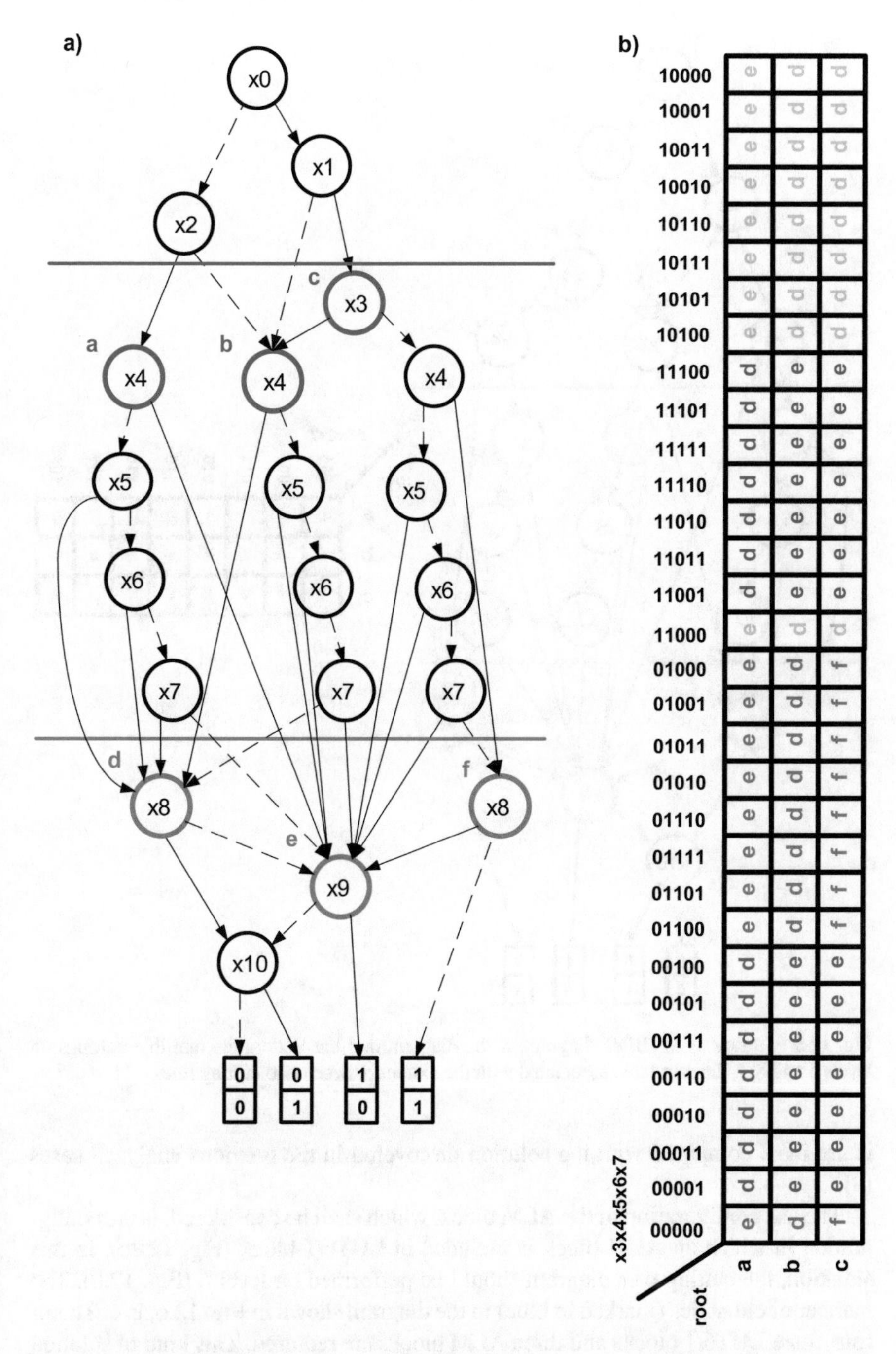

Fig. 12.4 Cutting of MTBDD diagram; **a** The diagram that has undergone multiple cuttings on levels 3 and 8, **b** The root table associated with the extract between two cutting lines

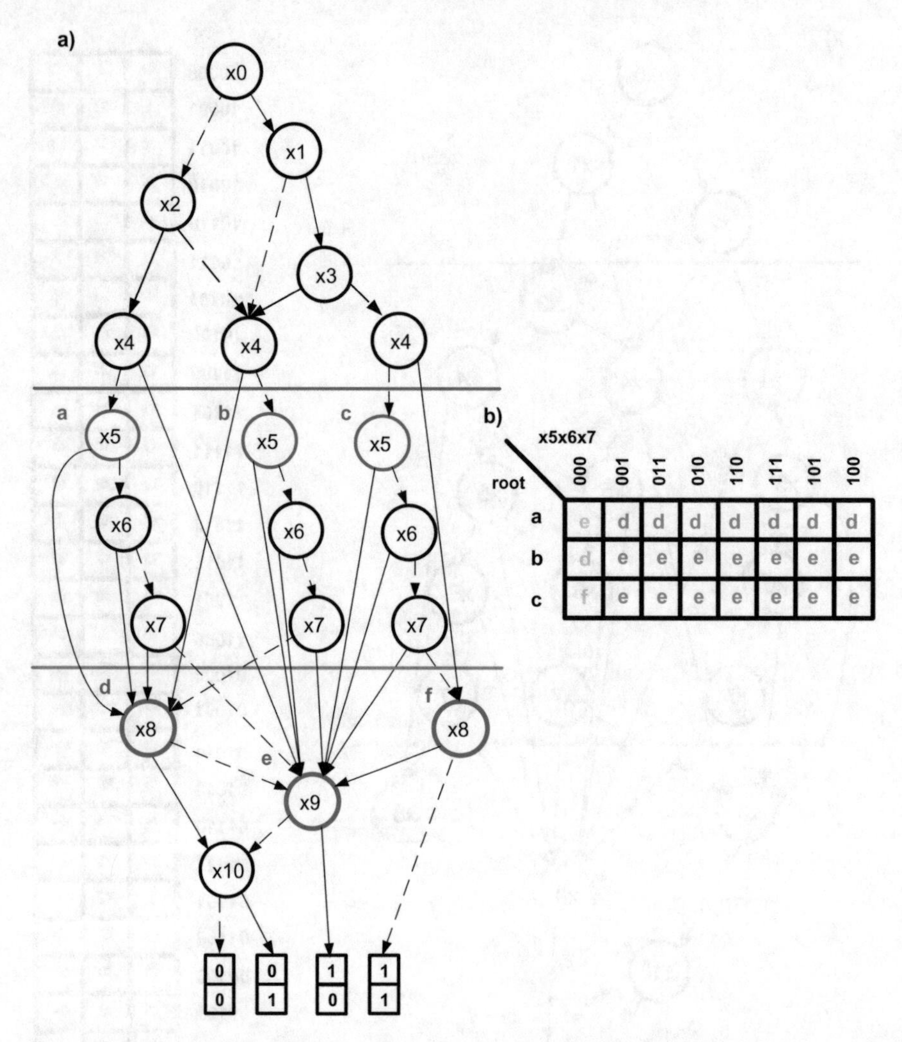

Fig. 12.5 Cutting of MTBDD diagram; **a** the diagram that has undergone multiple cuttings on levels 5 and 8, **b** the root table associated with the extract between two cutting lines

is not ideal compared with the solution discovered in the previous analyzed cases [9].

The last configuration of the ALM block, which shall be considered, is the configuration in which an ALM block is included in LUT6/1 block (Fig. 12.2c). In this situation, the cutting of a diagram should be performed on level 6 (Fig. 12.6). The number of cut nodes (marked in blue) in the diagram shown in Fig. 12.6, is 6. Therefore, three LUT6/1 blocks and three ALM blocks are required. This kind of solution is also worse than the previous solutions [9].

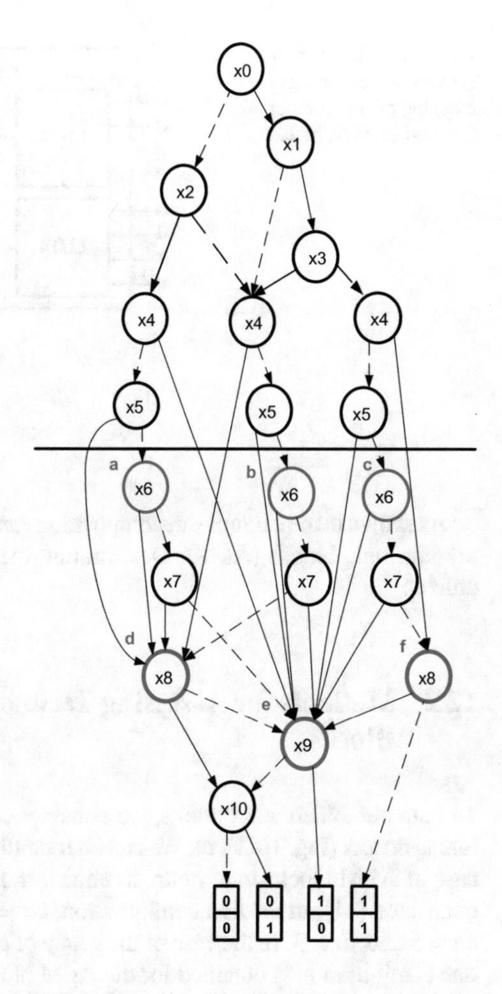

Fig. 12.6 MTBDD diagram that has undergone multiple cuttings on level 6

Among the analyzed cuttings of a diagram, which are matched to the configurable abilities of ALM blocks, the best result was obtained from the configuration of two independent LUT4/1 blocks. In this situation, bound blocks can be performed in a single ALM block. After replacing the extracts 0 and 1 with three bound functions (one extract is the variable x_0), the diagram that describes a free block, which is analyzed in the next stage of synthesis, includes six variables. Both bound functions, which are included in the multioutput function and describe a free block, depend on the same six variables. Thus, these blocks may be performed in a single ALM block that is configured in such a way that a part of the inputs is shared (4 in 6) for both LUT6/1 blocks [8].

As a result of the synthesis of the analyzed multioutput function using the multiple cutting method of an MTBDD diagram directed at technology mapping to ALM blocks, the obtained structure consists of two ALM blocks, as presented in Fig. 12.7 [9].

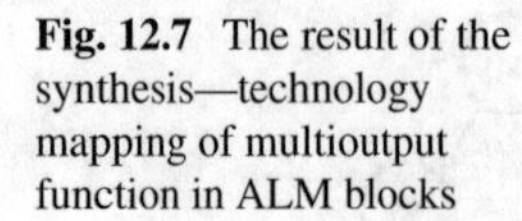

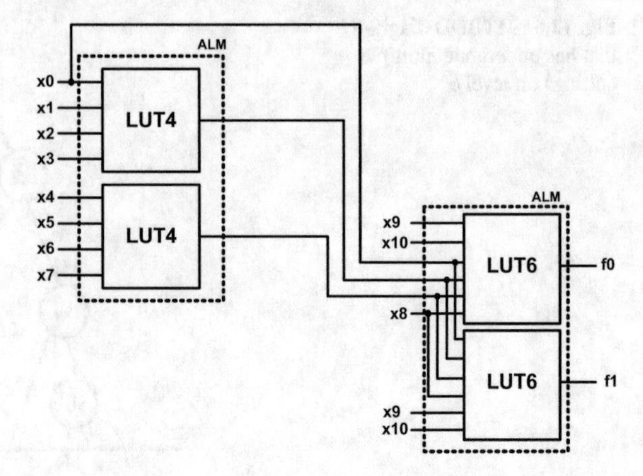

Fig. 12.7 The result of the synthesis—technology mapping of multioutput function in ALM blocks

As shown in the example, decomposition can focus on effective mapping in configuration logic blocks (ALM). Determining the effectiveness of mappings becomes crucial.

12.2 Methods for Assessing Technology Mapping in ALM Blocks

To map combinational circuits, we consider configurations in which no sharing of inputs occurs (Fig. 10.2a, c). We consider configurations in normal mode [11]. In the case of ALM blocks in a group, as shown in Fig. 10.2a, two configurations can be connected [11]. In the first configuration, $ka = kb = 4$; in the second configuration, $ka = 5$ and $kb = 3$. In the case of the group of configurations from Fig. 10.2b,c, only one configuration is obtained for the ALM blocks for which $k = 6$.

As presented in [9, 12], the technology mapping of combinational circuits in ALM blocks involves fitting bound blocks of a decomposed circuit into LUTs in configurable logic blocks. This mapping is shown in Fig. 12.8 [13]

The technology mapping shown in Fig. 12.8a is based on the choice of a cutting line to ensure that the number of variables located above the top cutting line *card(Xb1)* and the number of variables between the cutting lines *card(Xb2)* correspond to the number of inputs of the blocks LUTa (*ka*) and LUTb (*kb*). The mapping shown in Fig. 12.8b is based on a choice of the level of cutting to ensure that the number of bound variables *card(Xb)* is equal to (or at least lower than) the number of inputs of a single block LUT (k) included in a given configuration of an ALM block [13].

As illustrated in Fig. 12.8a, multiple decompositions performed using the multiple cutting method fits in the group of configurations from Fig. 10.2a. Decomposition performed using a single cutting fits in the group of configurations from Fig. 10.2c. The question of which group and which configuration will be the most effective in this case arises. This question is closely connected to the choice of a decomposition model

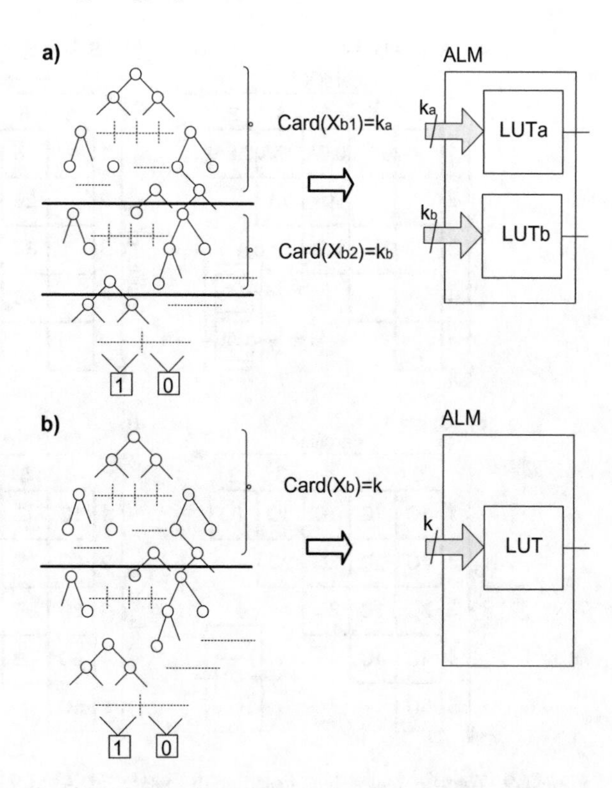

Fig. 12.8 Technology mapping of bound blocks: **a** for multiple decomposition performed using the multiple cutting method, and **b** for decomposition performed using a single cutting of BDD

and the cutting lines of the BDD. A given decomposition model can be described by the number of elements of a bound set (*card(Xb)*), on which the levels of cutting of BDD depend. The model can also be described by the number of bound functions (*numb_of_g*) that are associated with a given decomposition. The various methods for indicating the parameter *numb_of_g* are presented in [14, 15]. Thus, algorithms that enable us to choose a decomposition, such that the mappings in one of the configurations in ALM will be the most efficient taking into account the number of blocks should be developed [13].

We propose the use of triangle tables, as reported in previous literature [9, 13, 16], and modify these tables in such a way that the content of certain cells will describe the number of ALM blocks that are needed to implement the subcircuits that are obtained from the decompositions defined by the pair of parameters: *numb_of_g* and *card(Xb).* This idea is illustrated in Fig. 12.9 [13].

A triangle table, as shown in Fig. 12.9a, corresponds to the configuration *A*: $ka = kb = 4$. For two independent four-input LUTs, a block is configured such that the number of variables in a bound set cannot exceed four. The triangle table in Fig. 12.9a is partly filled because empty cells remain in the columns for which *card(Xb)* > 4. In the triangle tables, cases in which decomposition limits the number of variables are analyzed. Thus, the maximum number of bound functions introduced is *max(numb_of_g)* = *card(Xb)* – 1. In configuration *A*, a single ALM block may

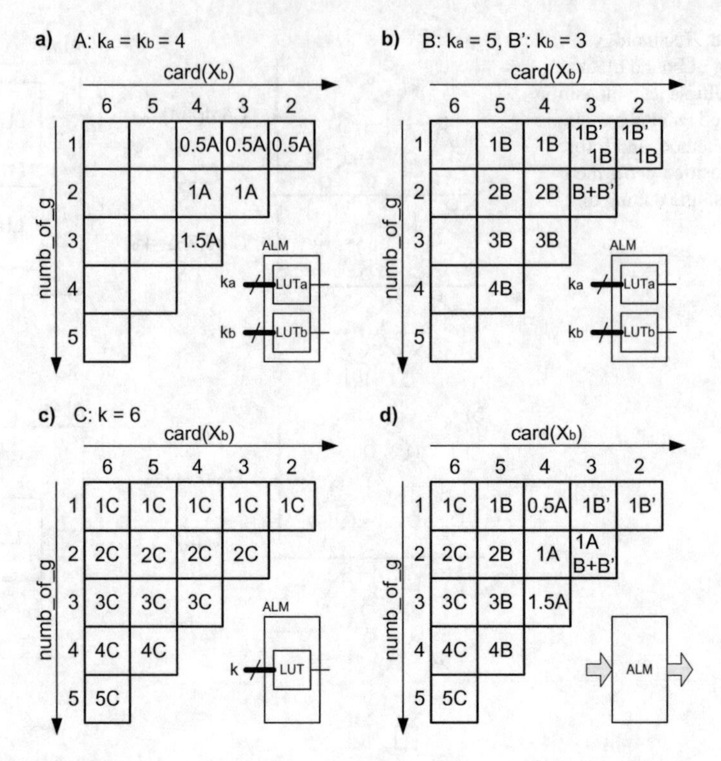

Fig. 12.9 Triangle tables that describe the usage of ALM blocks for **a** configuration A, **b** configuration B, **c** configuration C, and **d** a technology mapping table in ALM

implement two bound functions. The cells associated with the row *numb_of_g* = 2 contain the symbol *1A* (a single ALM block is needed in configuration *A*). The cells in rows *numb_of_g* = 1 and *numb_of_g* = 3 contain 0.5*A* (only one of two LUTs is utilized) and 1.5*A* symbols (three LUTs are necessary, i.e., 1.5 of ALM block) [13].

The triangle table in Fig. 12.9b corresponds to the configuration *ka* = 5 and *kb* = 3. For the pair *(card(Xb), numb_of_g),* we can use an LUT that has five inputs (marked in Fig. 12.9b as *B*), a LUT that has three inputs *(B')* or both LUTs *(B + B')*. A bound set with six bound variables is not used in this configuration; thus, the appropriate cells in the table remain empty. If a bound set has four or five elements, an LUT block that has five inputs (configuration *B*) is necessary. In other cases, for *numb_of_g* = 1, any LUT from a block may be employed (thus, *B* and *B'* are written in particular cells). However, when *numb_of_g* = 2, both blocks are employed to implement two bound functions (marked as *B* + *B'*) [13].

The triangle table in Fig. 12.9c corresponds to the configuration in which the ALM block implements a single function with six variables. In this case, the value in the cells in the table shown in Fig. 12.9c corresponds to the number of bound functions *numb_of_g*. This configuration is marked *C* [13].

Since each decomposition is accompanied by the pair of numbers *(card(Xb), numb_of_g)*, the issue of which decomposition best matches the ALM blocks or which ALM configuration (*A, B* or *C*) would be the most effective arises [13].

Based on an analysis of the tables from Fig. 12.9a–c, a results table can be created (a table of technology mapping in ALM) that will be used to choose the most efficient configuration. This results table is presented in Fig. 12.9d. When analyzing these results tables, for the cells associated with *card(Xb)* = 6, the only possibility is to choose configuration *C*. For *card(Xb)* = 5, configurations *C* or *B* may be chosen. Configuration *B* is a better choice since there a free LUT is associated with *B'*. In the case of *card(Xb)* = 4, configuration *C* is not effective, and thus, configurations *A* and *B* may be employed. Configuration *A* is a better solution because unused LUTs with four inputs (in configuration *B*, only LUTs with three inputs remain) can be applied. The described decomposition using a pair of numbers (2, 4) best corresponds with configuration *A*. In cases where *card(Xb)* < 4, the parameter *numb_of_g* is essential. When the value of this parameter is 1, configuration *B'* is the best choice because an LUT block with five inputs is available. When *numb_of_g* = 2, configurations *A* or *B* + *B'* may be chosen. In both cases, a single ALM block is employed and no free LUTs remain [13].

Based on an analysis of the technology mapping table in an ALM, a configuration can be chosen that would enable us to reduce as many ALM blocks as possible. For example, for a given function to undergo the decomposition described using the following pairs of numbers—(6,3), (5,2) or (4,1)—we need 3, 2 and 0.5 ALM blocks, respectively. In this case, the most effective decomposition is that described by the pair of numbers (4,1) [13].

Based on the analysis of a technology mapping table in ALM, an algorithm for choosing a configuration can be proposed, as shown in Algorithm 12.1.

Algorithm 12.1: Algorithm for choosing a configuration

```
1.  choose_a_configuration (f, set_of_available_ALM_configurations)
2.  {
3.  for(i = 0; i < set_of_available_ALM_configurations; i++)
4.  {
5.    decomposition = indicate_decomposition ( cutting_levels (i))
6.    numb_of_g = decomposition.numb_of_g;
7.    card(Xb) = decomposition.card(Xb);
8.        number_of_ALM   =   technology_mapping_table_ALM(numb_of_g,
    card(Xb))
9.  if( number_of_ALM < number_of_ALM_best)
10. {
11.   number_of_ALM _best = number_of_ALM;
12.   configuration_ALM_best = i;
13.   decomposition _best = decomposition
14. }
15. }
```

16. return(configuration_ALM_best, decomposition_best)
17. }

12.3 Technology Mapping of Nondisjoint Decomposition

The idea of implementing nondisjoint decomposition [17–19] is presented in Fig. 12.10a. When analyzing configurations of ALM blocks, in which sharing of inputs occurs, they correspond to some basic nondisjoint decompositions, as shown in Fig. 12.10b [13]. In ALM blocks, two configurations with sharing of inputs exist: *D:* $ka = 4$, $kb = 3$, $kab = 1$ and *E:* $ka = 3$, $kb = 3$, $kab = 2$. Configuration *D* can be associated with the decomposition in which *card(Xb)* = 4, *card(Xs)* = 1 and *card(Xf)* = 3 or *card(Xb)* = 3, *card(Xs)* = 1 and *card(Xf)* = 4. Configuration *E* can be associated with the decomposition in which *card(Xb)* = *card(Xf)* = 3 and *card(Xs)* = 2. One of the LUTs in an ALM block is connected with a bound block, and the other LUT is connected with a free block. The necessary condition in both configurations is the choice of non-disjoint decomposition, in which one bound function occurs. The choice of this decomposition requires feedback in an ALM block, which

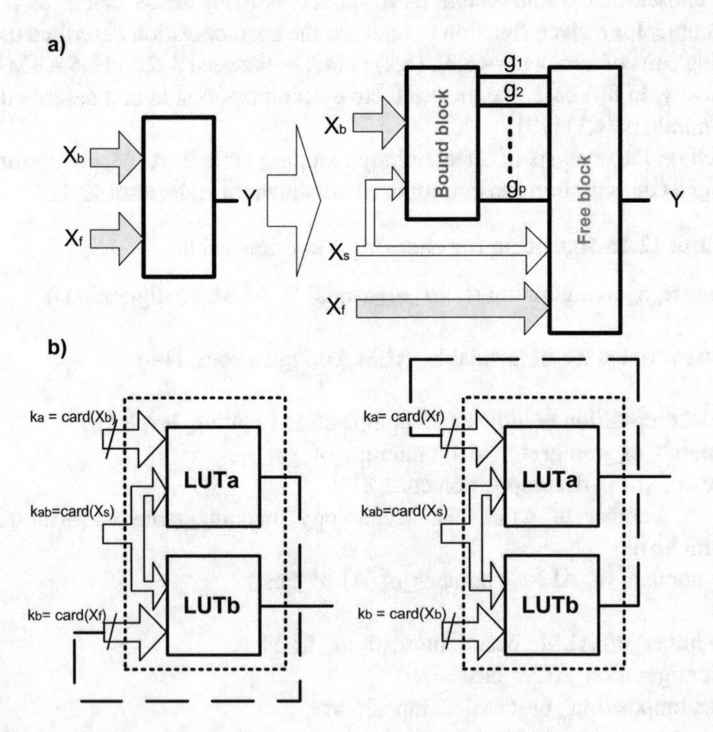

Fig. 12.10 Nondisjoint decomposition: **a** implementation, and **b** technology mapping in ALM blocks

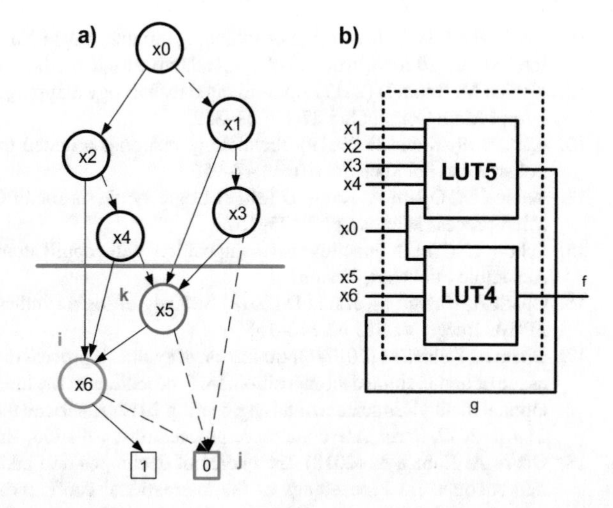

Fig. 12.11 Example of nondisjoint decomposition: **a** BDD, and **b** technology mapping in ALM blocks

may be problematic in some cases (limitations in subsequent stages of synthesis—placement and routing). Figure 12.11 presents an example of mapping of nondisjoint decomposition in an ALM block [13].

Configurations *D* and *E* are reasonable when implementing multioutput functions [14, 16].

References

1. Chen W, Zhang X, Yoshimura T, Nakamura Y (2011) A low power technology mapping method for adaptive logic module. In: International conference on field-programmable technology
2. Cong J, Hwang Y (2001) Boolean matching for LUT-based logic blocks with applications to architecture evaluation and technology mapping. IEEE Trans Comput Aided Des Integr Circuits Syst 20(9):1077–1090
3. Liang Y, Kuo T, Wang S. Mak W (2012) ALMmap: technology mapping for FPGAs with adaptive logic modules. IEEE Trans Comput-Aided Des Integr Circuit Syst 31(7):1134–1139
4. Lee KK, Wong DF (1999) An exact tree-based structural technology mapping algorithm for configurable logic blocks in FPGAs. In: IEEE international conference on computer design: VLSI in computers and processors, pp 1–6
5. Wu S, Hsu P, Mak W (2014) A novel wirelength-driven packing algorithm for FPGAs with adaptive logic modules. 19th Asia and South Pacific design automation conference (ASP-DAC) 2014, pp 501–506
6. Wu S-K, Hsu P-Y, Mak W-K (2014) A novel wirelength-driven packing algorithm for FPGAs with adaptive logic modules. In: 2014 19th Asia and South Pacific design automation conference (ASP-DAC), pp 501–506
7. Zhang X, Tu L, Chen D, Yuan Y, Huang K, Wang Z (2017) Parallel distributed arithmetic FIR filter design based on 4:2 compressors on Xilinx FPGAs. In: 4th international conference on signal processing, computing and control (ISPCC), pp 43–49
8. Altera (2012) Logic array blocks and adaptive logic modules in Stratix V Devices
9. Kubica M, Kania D (2017) Decomposition of multi-output functions oriented to configurability of logic blocks. Bull Polish Acad Sci Tech Sci 65(3):317–331

10. Curtis HA (1962) The design of switching circuits. D.van Nostrand Company Inc., Princeton
11. Intel Stratix 10 logic array blocks and adaptive logic modules user guide, UG-S10LAB, 2017
12. Kubica M, Kania D (2017) Area-oriented technology mapping for LUT-based logic blocks. Int J Appl Math Comput Sci 27(1):207–222
13. Kubica M, Kania D (2019) Technology mapping oriented to adaptive logic modules. Bull Polish Acad Sci Tech Sci 67(5):947–956
14. Kubica M, Opara A, Kania D (2017) Logic synthesis for FPGAs based on cutting of BDD. Microprocess Microsyst 52:173–187
15. Scholl C (2001) Functional decomposition with application to FPGA synthesis. Kluwer Academic Publisher, Boston
16. Opara A, Kubica M, Kania D (2018) Strategy of logic synthesis using MTBDD dedicated to FPGA. Integr: VLSI J 62:142–158
17. Opara A, Kubica M (2017) Optimization of synthesis process directed at FPGA circuits with the usage of non-disjoint decomposition. In: Proceedings of the international conference of computational methods in sciences and engineering 2017. American Institute of Physics, Thessaloniki, 21 Apr 2017, Seria: AIP conference proceedings; vol 1906, Art. no. 120004
18. Opara A, Kubica M (2018) The choice of decomposition taking non-disjoint decomposition into account. In: Proceedings of the international conference of computational methods in sciences and engineering 2018. American Institute of Physics, Thessaloniki, 14 Mar 2018, Seria: AIP conference proceedings; vol 2040, Art. no. 080010
19. Opara A, Kubica M, Kania D (2019) Methods of improving time efficiency of decomposition dedicated at FPGA structures and using BDD in the process of cyber-physical synthesis. IEEE Access 7:20619–20631

Chapter 13
Decomposition Methods of FSM Implementation

In the group of digital circuits, we can distinguish between combinational circuits and sequential circuits. The mathematical model of the sequential system is an FSM. An FSM consists of a memory block and combination blocks that are responsible for controlling memory elements and setting output states. The process of synthesis of sequential systems consists of numerous stages, among which coding of an internal FSM states and the synthesis of excitation functions and output functions can be distinguished [1–3]. A certain number of different algorithms for coding internal states are known [4, 5. 2]. The process of coding internal states is directed at specific goals, i.e., minimizing the area of an FSM [6–11], speed of an FSM [12], and power consumption [13–20]. Some studies focused on specific architectures of PLD devices [8, 21–23] or even the specific resources of these devices [24, 25]. If LUT-based FPGAs are employed, the synthesis algorithms described in previous chapters can be used to implement FSM combination blocks. Some additional optimization elements are presented in this chapter.

13.1 Finite State Machine Background

An FSM is an abstract mathematical model of a sequential circuit. This model is the five-tuple $\{X,Y,S,\delta,\lambda\}$, where X is a set of input vectors, S is the set of internal states, and Y is the set of output vectors. The relations among these elements are determined by a transition function, which is usually marked with the symbol δ and an output function (λ) that describes the values of output signals. A general structure of an FSM is shown in Fig. 13.1. Transition functions and output functions are associated with combinational blocks that are marked as δ and λ, respectively. A memory block was symbolically marked as a D-type Flip-Flop (DFF) in Fig. 13.1.

In accordance with the designations, included in Fig. 13.1, a transition function is determined by $\delta: B^{N+K} \rightarrow B^K$, where $B = \{0,1\}$. Depending on the type of automaton $\lambda: B^K \rightarrow B^M$ (Moore's automaton) or $\lambda: B^{N+K} \rightarrow B^M$ (Mealy's automaton). In the

M. Kubica et al., *Technology Mapping for LUT-Based FPGA*, Lecture Notes in Electrical Engineering 713, https://doi.org/10.1007/978-3-030-60488-2_13

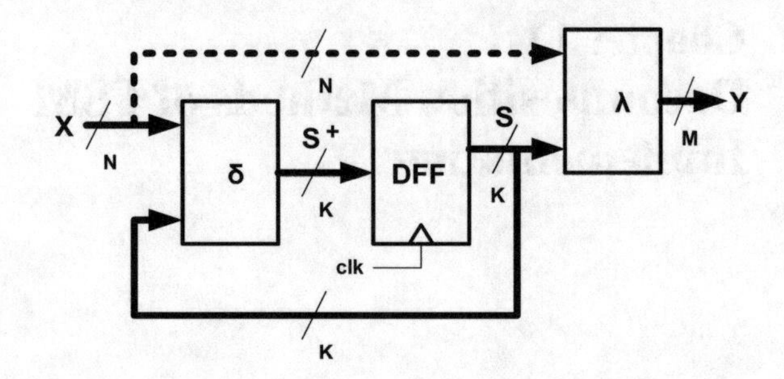

Fig. 13.1 Block scheme of FSM

process of coding states, each internal state of an automaton is associated with the K-bit vector of a state. The number of a state vector's bits depends on the accepted method of coding internal states and fulfills the inequality $lg_2(card(S) \leq K \leq card(S)$ [21].

13.2 Technology Mapping of the FSM Combination Part in Configuration Logic Blocks

We consider the FSM technology mapping in configuration logic blocks in the example of ALM blocks [26]. The mapping process will use the sharing of logic resources between the transition function δ and the output function λ. We consider an exemplary benchmark referred to as beecount [27]. This automaton has three inputs ($n = 3$), four outputs ($m = 4$) and seven states ($card(S) = 7$). The number of bits needed to code seven states is $k = lg_2(card(S)) = 3$. Beecount is Mealy's automaton. In addition to a three-bit register ($k = 3$), two combinational blocks that are described by the transition function $\delta: B^{n+k} \rightarrow B^k$ and the output function $\lambda: B^{n+k} \rightarrow B^m$ can be distinguished. To implement a transition block and an output block based on ALM blocks, seven blocks can be utilized, in which the entire ALM block implements only one six-input function. Three ALM blocks can be used to implement a transition function ($\delta: B^6 \rightarrow B^3$), and the remaining blocks can be used to implement the output function ($\lambda: B^6 \rightarrow B^4$) [10].

The question of whether another way to configure ALM blocks and efficiently use the resources of an FPGA structure arises.

The search for an effective mapping starts with the decomposition of a transition function and an output function, which belong to many different decompositions. The following three decompositions can be distinguished:

1. $\delta: X \times S \rightarrow S^+ \Rightarrow \delta: \{i2,i1,i0,Q2,Q1,Q0\} \rightarrow \{q2, q1, q0\}$

$$\delta_1: \{i2, i1, i0, Q2, Q1, Q0\} \rightarrow \{q2\}$$
$$X_b = \{i2, i1, Q2, Q1, Q0\}; \quad X_f = \{i0\};$$

$$v(\text{i0}|\text{i2, i1, Q2, Q1, Q0}) = 2 \qquad \Rightarrow numb\_of\_g = 1$$

$$\delta_2 : \{\text{i2, i1, i0, Q2, Q1, Q0}\} \rightarrow \{\text{q1,q0}\}$$
$$X_b = \{\text{i1, Q2, Q1, Q0}\}; \quad X_f = \{\text{i2,i0}\};$$
$$v(\text{i2, i0}|\text{i1, Q2, Q1, Q0}) = 4 \qquad \Rightarrow numb\_of\_g = 2$$

2. λ: X × S → Y ⇒ λ: {i2,i1,i0,Q2,Q1,Q0}→{ o3,o2,o1,o0}

$$X_b = \{\text{i2, i1, i0, Q2, Q1}\}; \quad X_f = \{\text{Q0}\};$$
$$v(\text{Q0}|\text{i2, i1, i0, Q2, Q1}) = 8; \qquad \Rightarrow numb\_of\_g = 3$$

Finding and choosing these decompositions enables an effective mapping of an FSM in ALM blocks, as illustrated in Fig. 13.2 [10, 28, 29].

The technology mapping presented in Fig. 13.2 uses four ALM blocks that work in a configuration with one shared input (LUT5/1 + LUT4/1 – one common input) and two ALM blocks in a configuration without sharing of inputs (LUT4/1 + LUT4/1). An LUT2/1 block that can be implemented in each configuration is necessary; resources can be used in other ways [10].

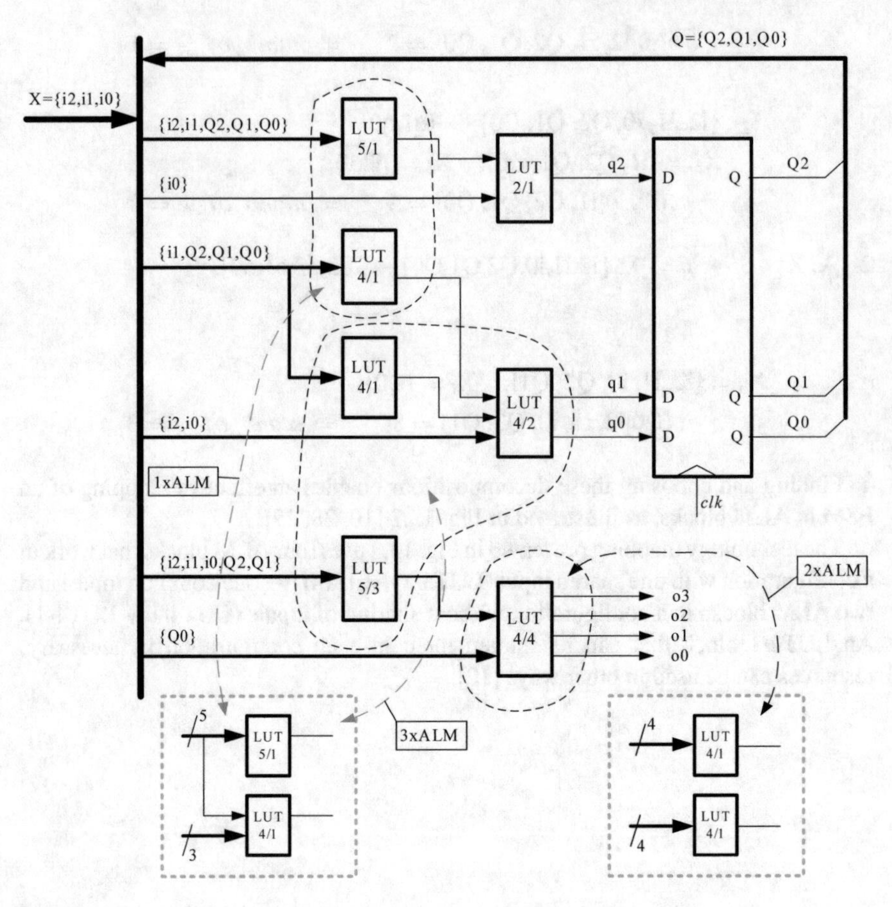

Fig. 13.2 Technology mapping of the beecount automaton in ALM blocks

References

1. Lin Y, Gang Q, Villa T, Sangiovanni-Vincentelli A (2008) An FSM reengineering approach to sequential circuit synthesis by state splitting. IEEE Trans Comput Aided Des Integr Circuits Syst 27(6):1159–1164
2. Villa T, Sangiovanni-Vincentelli A (1990) NOVA: state assignment for finite state machines for optimal two-level logic implementation. IEEE Trans Comput-Aided Des 9(1990):905–924
3. Yang S, Ciesielski M (1991) Optimum and suboptimum algorithms for input encoding and its relationship to logic minimization. IEEE Trans Comput Aided Des Integr Circuits Syst 10:4–12
4. Deniziak S, Wisniewski M (2010) FPGA-based state encoding using symbolic functional decomposition. Electron Lett 46(19):1316–1318
5. Devadas S, Newton AR, Ashar P (1991) Exact algorithms for output encoding, state assignment and four-level Boolean minimization. IEEE Trans Comput-Aided Des 10:13–27
6. Barkalov A, Titarenko L, Mielcarek K (2018) Hardware reduction for LUT-based Mealey FSMs. Int J Appl Math Comput Sci 28(3):595–607
7. Czerwiński R, Kania D (2012) Area and speed oriented synthesis of FSMs for PAL-based CPLDs. Microproces Microsyst 36(1):45–61

8. Czerwiński R, Kania D (2009) CPLD–oriented synthesis of finite state machines. In: Proceedings of the twelfth Euromicro symposium on digital system design. In: DSD2009, IEEE Computer Society Press, Patras, pp 521–528
9. Czerwiński R, Kania D (2009) Synthesis of finite state machines for CPLDs. Int J Appl Math Comput Sci (AMCS) 19(4):647–659
10. Kubica M, Kania D (2019) Technology mapping oriented to adaptive logic modules. Bull Polish Acad Sci Tech Sci 67(5):947–956
11. Kubica M, Kajstura K, Kania D (2018) Logic synthesis of low power FSM dedicated into LUT-based FPGA. In: Proceedings of the international conference of computational methods in sciences and engineering 2018. In: American Institute of Physics, Thessaloniki, 14 Mar 2018, Seria: AIP conference proceedings, vol 2040
12. Czerwiński R, Kania D (2010) A synthesis of high speed finite state machines. Bull Polish Acad Sci Tech Sci 58(4):635–644
13. Benini L, DeMicheli G (1995) State assignment for low power dissipation. IEEE J Solid-State Circuit 30(3):258–268
14. Grzes T, Solov'ev V (2014) Sequential algorithm for low-power encoding internal states of finite state machines. Int J Comput Syst Sci 53(1):92–99
15. Kajstura K, Kania D (2011) A decomposition state assignment method of finite state machines oriented towards minimization of Power. Przegląd Elektrotechniczny 87(6):146–150
16. Kajstura K, Kania D (2016) Binary tree-based low power state assignment algorithm. In: 12th international conference of computational methods in science and engineering, ICCMSE 2016, 17–20 Mar 2016, Athens, Greece, AIP conference proceedings 1790, pp 0300007_1–0300007_4
17. Kajstura K, Kania D (2018) Low power synthesis of finite state machines state assignment decomposition algorithm. J Circuit Syst Comput 27(3):1850041–1–1850041–14
18. Mengibar L, Entrena L, Lorenz MG, Millan ES (2005) Partitioned state encoding for low Power in FPGAs. Electron Lett 41:948–949
19. Nawrot R, Kulisz J, Kania D (2017) Synthesis of energy-efficient FSMs implemented in PLD circuits. In: 13th international conference of computational methods in science and engineering, ICCMSE 2017, 21–25 Apr 2017, Thessaloniki, Greece, AIP conference proceedings 1906, pp 120003_1–120003_4
20. Venkataraman G, Reddy SM, Pomeranz I (2003) GALLOP: genetic algorithm based low power FSM synthesis by simultaneous partitioning and state assignment. In: 16th international conference on VLSI design, pp 533–538
21. Czerwiński R, Kania D (2013) Finite state machine logic synthesis for CPLDs. Springer, Berlin, vol 231, XVI
22. Kaviani A, Brown S (2000) Technology mapping issues for an FPGA with lookup tables and PLA-like blocks. In: Proceedings of the 2000 ACM/SIGDA eighth international symposium on field programmable gate arrays, pp 60–66
23. Wisniewski R, Barkalov A, Titarenko L, Halang W (2011) Design of microprogrammed controllers to be implemented in FPGAs. Int J Appl Math Comput Sci 21(2):401–412
24. Czerwinski R, Kania D (2016) State assignment and optimization of ultra high speed FSMs utilizing tri-state buffers. ACM Trans Des Autom Electron Syst 22(1):1–25
25. Kulisz J, Nawrot R, Kania D (2016) Synthesis of energy-efficient counters implemented in PLD circuits. In: 12th international conference of computational methods in science and engineering, ICCMSE 2016, 17–20 Mar 2016, Athens, Greece, AIP conference proceedings 1790, pp 0300006_1–0300006_4
26. Altera (2012) Logic array blocks and adaptive logic modules in Stratix V Devices
27. Collaborative Benchmarking Laboratory, Department of Computer Science at North Carolina State University, https://www.cbl.ncsu/edu/
28. Kubica M, Kania D (2017) Decomposition of multi-output functions oriented to configurability of logic blocks. Bull Polish Acad Sci Tech Sci 65(3):317–331
29. Opara A, Kubica M, Kania D (2018) Strategy of logic synthesis using MTBDD dedicated to FPGA. Integr VLSI J 62:142–158

Chapter 14
Algorithms for Decomposition and Technological Mapping

The methods described in this book lead to two synthesis strategies: dekBDD [1, 2] and MultiDec [1, 2]. The first one uses single cutting line for decomposition, and the second one multiple cuttings of BDD for decomposition. DekBDD and MultiDec strategies are appropriately described by Algorithm 1, and Algorithm 3 [2].

Figure 14.1a shows the dekBDD algorithm, where one of the parameters is card (X_b). In the case of mapping the circuit in k-input LUT blocks, this parameter can be changed in the range from 2 to k, leading to the most effective technology mapping based on the analysis of triangle tables.

The essence of Algorithm 1 is the implementation of a double loop using indexes i and j. Within these loops the following steps are performed:

- swap of x_i and x_j variables,
- the use of merging and splitting (Algorithm 2) to determine the effective set of functions,
- searching for a nondisjoint decomposition [2, 3].

The key element of the algorithm is determining a cofactor of the mapping δ_{tmp}. Its value is compared with the best solution so far (δ). If the solution obtained in a given step is better, it is remembered.

The algorithm of quick merging and splitting functions is shown in Fig. 14.1b. The essence of algorithm 2 is to create PMTBDD diagrams so that it does not lead to increase in the number of cut nodes for a given variable ordering. The α parameter was introduced to prevent merging too many functions, which may lead to inefficient solutions. The set of functions F: f_0, …, f_{m-1} is sorted ascending according to the number of cut nodes. Then, for a single function or set of functions F[i] an attempt is made to attach another function F[j]. The number of cut nodes is determined for the resulting T diagram. The rounded number of cut nodes (RoundPow2CutNodes) determines whether the combination of the functions F[i] and F[j] is acceptable. If the process of attaching is accepted, F[i] is removed from the list and F[j] is replaced with the newly created group T. In the next steps, the next functions are attempted to be included in the set [2].

M. Kubica et al., *Technology Mapping for LUT-Based FPGA*, Lecture Notes in Electrical Engineering 713, https://doi.org/10.1007/978-3-030-60488-2_14

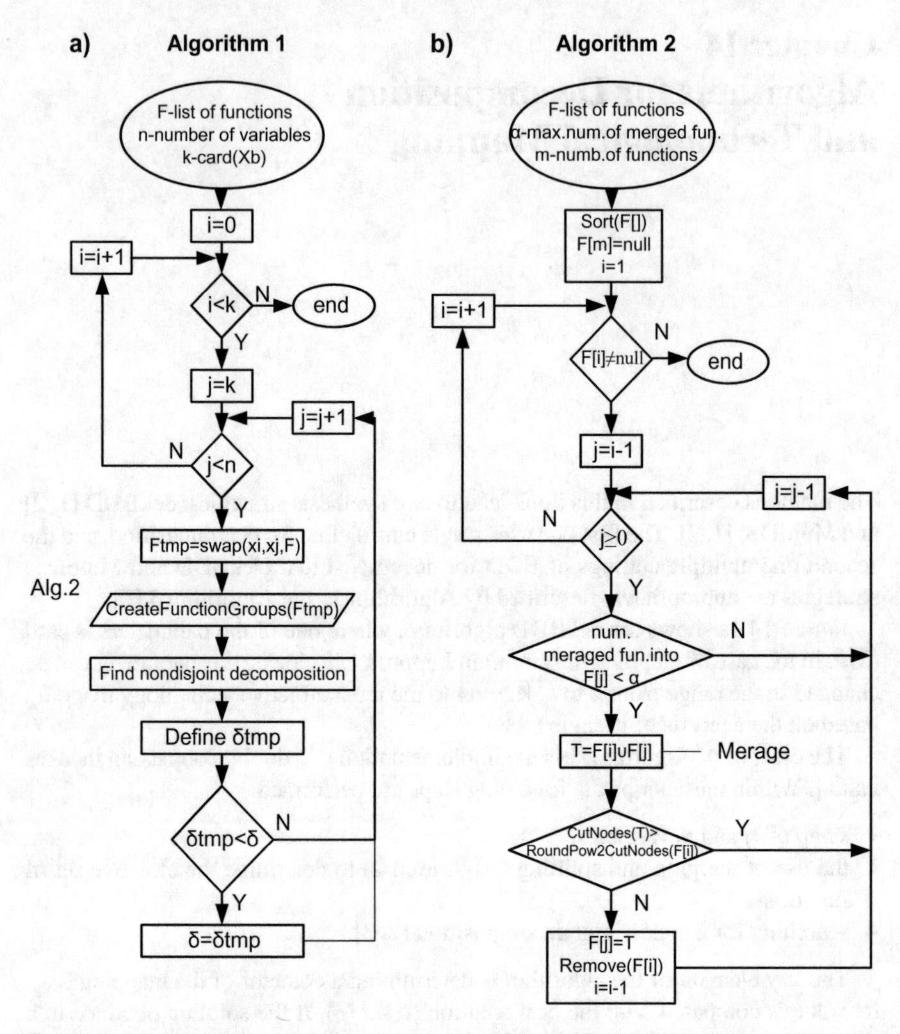

Fig. 14.1 A block scheme: **a** decomposition algorithm 1 (dekBDD) of multioutput function, **b** Algorithm 2 of quick merging and splitting functions[1] [2]

Figure 14.2 illustrates Algorithm 3, which describes a MultiDec strategy.

The MultiDec algorithm (Algorithm 3) is similar to the dekBDD algorithm (Algorithm 1). A characteristic feature of the implementation of decomposition by the method of multiple cutting is that for each variable x_i there is a set of acceptable positions in BDD (variable ID set). Depending on the set in which a given variable is to be compact, the variable x_i is assigned to a specific position. The variable p is used to specify this position. In the case of algorithm 3, δ_{tmp} is determined by taking into account all bound blocks.

---

[1] © (2020) IEEE. Reprinted with permission from Opara et al. [2]

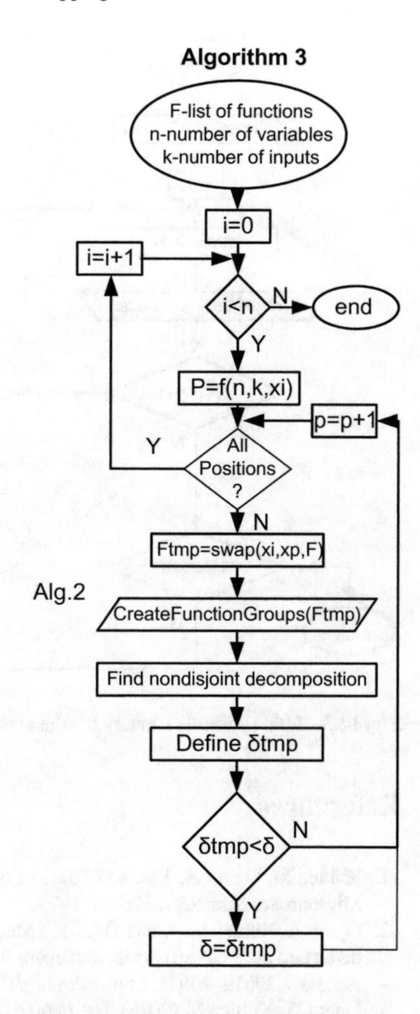

Fig. 14.2 A block scheme: decomposition algorithm 3 (MultiDec) of a multioutput function (see Footnote 1) [2]

Both Algorithms (1 and 3) operate on a single cluster. Separate functions must be partitioned into separate clusters and an initial variable ordering must be taken. If the results for both algorithms are not obtained, it becomes necessary to split the cluster, and if that fails, decomposition with Shannon's expansion is carried out. The methodology is shown in the form of algorithm 4 and presented in Fig. 14.3 [2].

Both algorithms have been implemented as tools with corresponding names. Based on the input descriptions of .pla, these tools can determine the number of necessary logic blocks, levels and synthesis time. In addition, tools can generate HDL description of decomposed functions that can be further synthesized in other tools. Additional information about the subject of the synthesis algorithms is provided in [2, 4–10].

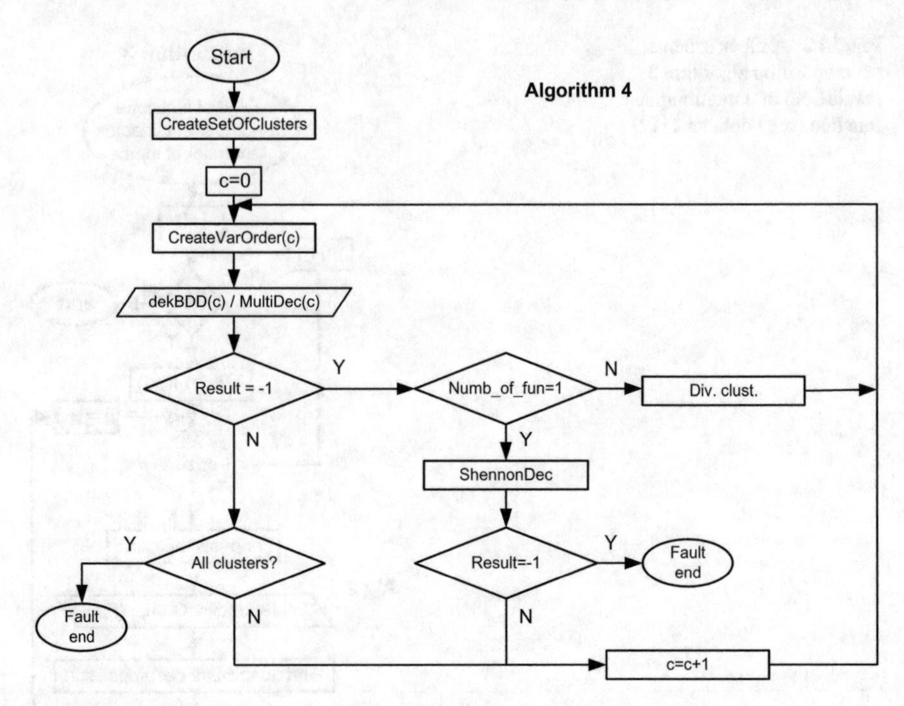

Fig. 14.3 Total synthesis strategy for the algorithms dekBDD and MultiDec (see Footnote 1) [2]

References

1. Kubica M, Opara A, Kania D (2017) Logic synthesis for FPGAs based on cutting of BDD. Microproces Microsyst 52:173–187
2. Opara A, Kubica M, Kania D (2019) Methods of improving time efficiency of decomposition dedicated at FPGA structures and using BDD in the process of cyber-physical synthesis. IEEE Access 7:20619–20631. https://doi.org/10.1109/ACCESS.2019.2898230
3. Opara A, Kubica M (2018) The choice of decomposition taking non-disjoint decomposition into account. In: Proceedings of the international conference of computational methods in sciences and engineering 2018. American Institute of Physics, Thessaloniki, 14 Mar 2018, Seria: AIP conference proceedings; vol 2040, Art. no. 080010
4. Kubica M, Kania D (2015) New concept of graph for function decomposition, programmable devices and embedded systems. IFAC conference on PDES 2015, pp 61–66
5. Kubica M, Kania D, Opara A (2016) Decomposition time effectiveness for various synthesis strategies dedicated to FPGA structures. 12th international conference of computational methods in science and engineering, ICCMSE 2016, 17–20 March 2016, Athens, Greece, AIP conference proceedings 1790, pp 0300005_1–0300005_4
6. Kubica M, Kania D (2017) Area-oriented technology mapping for LUT-based logic blocks. Int J Appl Math Comput Sci 27(1):207–222
7. Kubica M, Kania D (2017) Decomposition of multi-output functions oriented to configurability of logic blocks. Bull Polish Acad Sci Tech Sci 65(3):317–331
8. Kubica M, Kania D (2019) Technology mapping oriented to adaptive logic modules. Bull Polish Acad Sci Tech Sci 67(5):947–956

9. Opara A, Kubica M (2017) Optimization of synthesis process directed at FPGA circuits with the usage of non-disjoint decomposition. In: Proceedings of the international conference of computational methods in sciences and engineering 2017. American Institute of Physics, Thessaloniki, 21 Apr 2017, Seria: AIP conference proceedings; vol 1906, Art. no. 120004
10. Opara A, Kubica M, Kania D (2018) Strategy of logic synthesis using MTBDD dedicated to FPGA. Integr VLSI J 62:142–158

Chapter 15
Results of Experiments

To prove the efficiency of the proposed algorithms, a series of experiments has been conducted. Popular benchmarks [1], which describe combinational circuits in the pla format, underwent experiments. Depending on the series of experiments, the results may vary slightly due to the settings of the synthesis tools.

15.1 Comparison of MultiDec and DekBDD Systems

In Table 15.1 the synthesis results for dekBDD and MultiDec systems for Spartan 3 blocks [2] are presented. This table contains the following columns: “Blocks”—describing the number of necessary logic blocks (number of inputs k = 4 or k = 5), “Levels”—the number of logical levels for the critical path, “g_disjoint”—number of disjoint bound functions, “G_nondisjoint”—number of nondisjoint bound functions. In addition, the table specifies: “g_non/g_dis”—relation number of nondisjoint bound functions to number of disjoint bound functions, “bdd swap”—number of calls to bdd_swap functions and “time”—synthesis time [3].

Analyzing the result presented in Table 15.1 it can be seen that in many cases decomposition in the MultiDec system fails. This is due to memory usage requirements. The last rows of the table contain the sum of individual results. In addition, the sum value was also placed only for benchmarks for which results were obtained in both tools (dekBDD v MultiDec).

Comparing the results, we can see that the dekBDD system gives better results in terms of the number of logic blocks, while the MultiDec system is better for the number of logic levels. It is worth noting that the dekBDD system is much more effective in searching for nondisjoint decomposition, which leads to a reduction in the number of logical blocks used. Considering the dekBDD system, for circuits where the number of blocks exceeds 300, the number of nondisjoint variables has a maximum of 36% of the number of functions g. By analyzing the number of calls to the bdd_swap function you can see that this number is higher for the dekBDD system.

M. Kubica et al., *Technology Mapping for LUT-Based FPGA*, Lecture Notes in Electrical Engineering 713, https://doi.org/10.1007/978-3-030-60488-2_15

Table 15.1 Comparison of dekBDD with MultiDec (© (2020) IEEE. Reprinted with permission from Opara et al. [3])

Benchmarks			dekBDD						
Name	In	Out	Blocks	Levels	g_disjoint	g_nondis	g_non/g_dis	bdd swaps	time [s]
5xp1	7	10	9.5	2	3	3	1.000	1526	0.01
9sym	9	1	5.5	5	8	3	0.375	1004	0.01
alu2	10	8	17.5	4	16	14	0.875	7262	0.02
alu4	14	8	352	18	396	66	0.167	73,971	0.65
apex3	54	50	2311	42	2581	478	0.185	638,680	15.98
apex7	49	37	62	8	44	99	2.250	97,083	0.85
b12	15	9	13.5	2	7	5	0.714	5962	0.03
b9	16	5	34	7	39	34	0.872	16,818	0.08
c8	28	18	21	4	10	15	1.500	10,684	0.05
cht	47	36	23	2	1	0	0.000	3760	0.01
clip	9	5	18	5	14	17	1.214	8372	0.07
cm162a	14	5	8	4	6	2	0.333	4148	0.03
cm163a	16	5	7.5	3	4	2	0.500	2700	0.01
cm85a	11	3	6.5	3	5	1	0.200	3328	0.03
con1	7	2	2.5	2	1	1	1.000	794	0
count	35	16	31	6	22	32	1.455	34,358	0.17
duke2	22	29	186	12	179	157	0.877	139,333	1.36
e64	65	64	78	16	38	0	0.000	692,924	3.99
ex1010	10	10	573	8	624	41	0.066	117,212	0.68
example2	85	66	87.5	7	52	83	1.596	189,532	0.81
f51m	8	8	7	3	2	2	1.000	1276	0.01
inc	7	9	15	3	11	6	0.545	4222	0.01
misex1	8	7	9.5	3	6	10	1.667	3534	0.02
misex2	25	18	27	3	17	6	0.353	12,044	0.05
misex3	14	14	251.5	13	268	98	0.366	68,276	0.86
misex3c	14	14	103	10	100	53	0.530	31,578	0.23
mux	21	1	15.5	9	16	20	1.250	5574	0.03
pcle	19	9	15.5	4	9	11	1.222	12,524	0.04
pdc	16	40	83.5	7	60	58	0.967	53,927	0.69
rd73	7	3	4.5	2	3	0	0.000	690	0.01
rd84	8	4	7.5	3	5	1	0.200	1464	0.01
sao2	10	4	25	5	23	14	0.609	6928	0.04
sct	19	15	15	4	5	14	2.800	8040	0.03
spla	16	46	259.5	11	243	122	0.502	199,556	1.94

(continued)

Table 15.1 (continued)

Benchmarks			dekBDD						
Name	In	Out	Blocks	Levels	g_disjoint	g_nondis	g_non/g_dis	bdd swaps	time [s]
t481	16	1	4.5	4	4	2	0.500	1634	0.03
table 3	14	14	643	16	730	94	0.129	138,116	1.67
term1	34	10	21.5	7	16	28	1.750	21,004	0.21
ttt2	24	21	27.5	3	15	20	1.333	12,376	0.07
vda	17	39	632.5	14	689	249	0.361	213,883	3.1
vg2	25	8	44	8	44	61	1.386	35,651	0.25
x1	51	35	95	9	75	85	1.133	51,174	0.27
x2	10	7	9.5	3	6	10	1.667	2488	0.01
x4	94	71	77.5	5	28	92	3.286	216,384	0.66
Z5xp1	7	10	9.5	2	3	3	1.000	1526	0.01
z9sym	9	1	5.5	5	8	3	0.375	1004	0.02
Sum:			6256	316	6436	2115	0.329	3,154,324	35.11
dekBDD v MultiDec			181.5	61	114	95	0.833	59,572	0.31
Benchmarks			MultiDec						
Name	In	Out	Blocks	Levels	g_disjoint	g_nondis	g_non/g_dis	bdd swaps	time [s]
5xp1	7	10	14	2	8	1	0.125	900	0.16
9sym	9	1	4.5	3	6	2	0.333	479	0.48
alu2	10	8	21.5	2	19	5	0.263	1952	2.79
alu4	14	8							
apex3	54	50							
apex7	49	37							
b12	15	9	22.5	3	16	9	0.563	2607	0.37
b9	16	5							
c8	28	18							
cht	47	36	23	2	1	1	1.000	128	0.97
clip	9	5							
cm162a	14	5							
cm163a	16	5							
cm85a	11	3	15	3	14	8	0.571	1755	2.32
con1	7	2	2.5	2	1	1	1.000	129	0.11
count	35	16							
duke2	22	29							

(continued)

Table 15.1 (continued)

Benchmarks			MultiDec							
Name	In	Out	Blocks	Levels	g_disjoint	g_nondis	g_non/g_dis	bdd swaps	time [s]
e64	65	64							
ex1010	10	10							
example2	85	66							
f51m	8	8	15.5	4	11	3	0.273	1323	1.20
inc	7	9	21	3	14	5	0.357	1388	1.39
misex1	8	7	10.5	3	7	3	0.429	1321	1.50
misex2	25	18							
misex3	14	14							
misex3c	14	14							
mux	21	1							
pcle	19	9	11.5	3	6	3	0.500	16190	13.12
pdc	16	40							
rd73	7	3	4.5	2	3	0	0.000	537	0.51
rd84	8	4	8	2	7	0	0.000	928	0.15
sao2	10	4							
sct	19	15							
spla	16	46							
t481	16	1	3.5	3	6	6	1.000	1212	1.75
table 3	14	14							
term1	34	10							
ttt2	24	21							
vda	17	39							
vg2	25	8							
x1	51	35							
x2	10	7	12.5	3	9	3	0.333	1000	1.22
x4	94	71							
Z5xp1	7	10	14	2	8	1	0.125	895	0.18
z9sym	9	1	4.5	3	6	2	0.333	479	0.48
Sum:			208.5	45	142	53	0.373	33223	28.70
dekBDD v MultiDec			208.5	45	142	53	0.373	33223	28.70

Since the synthesis time is much longer for the MultiDec system, it can be stated that the number of bdd_swap operations does not significantly affect the synthesis time. Much worse results in terms of synthesis time for MultiDec are due to poor time optimization of this tool (creating and coping with additional multidimensional tables and other structures). Determining the dependency of the synthesis time from

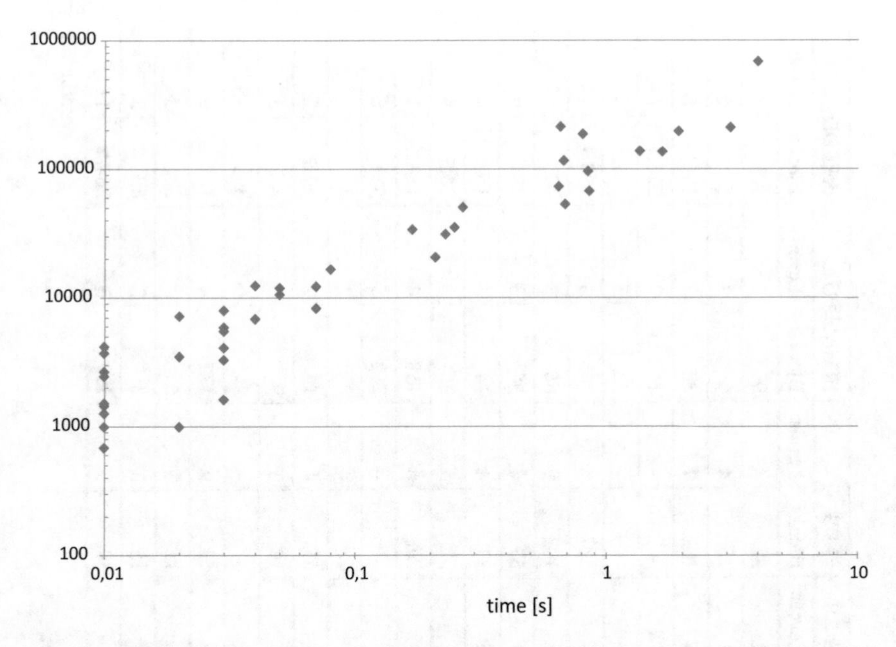

Fig. 15.1 Dependency of the number of bdd_swap and synthesis time for dekBDD (© (2020) IEEE. Reprinted with permission from Opara et al. [3])

the number of calls of a swap function is important, as illustrated in the form of the graph in Fig. 15.1 [3].

The dependency of synthesis time and the number of operations that change the sequence of variables (bdd_swap ()) is presented in Fig. 15.1. Considering this relationship for the dekBDD system, it can be seen that it has a linear character. For small processing times, the points on the graph are grouped into vertical lines due to the small resolution of measuring time, which is limited by the coating of an operating system. A more accurate comparison of both systems can be found in [3].

15.2 Comparison with Selected Academic Systems

The system dekBDD enables the decomposition process to be carried out for a higher number of benchmarks and produces better solutions than MultiDec with respect to the number of logic blocks LUT4 or LUT5 (LUT45). The results obtained for dekBDD will be compared with the solutions obtained by other academic tools. Direct comparisons of different academic tools are presented in Table 15.2[1] [4].

The results for the FLDS [5] are given for LUT4/1. In other cases, the results are presented for the LUT with 5 inputs. In addition to the results for the FLDS,

[1] © Reprinted from Kubica et al. [4], Copyright (2020), with permission from Elsevier.

Table 15.2 Comparison of academic tools with dekBDD (see Footnote 1) [4]

Benchmarks			DAOmap	DDBDD		ABC		BDS - PGA 2.0		FLDS		IRMA2FPGA		dekBDD	
Name	In	Out	Blocks	Blocks	Levels	Blocks	Levels	Blocks	Levels	Blocks	Levels	Blocks	Levels	Blocks	Levels
5xp1	7	10	–	19	2	23	3	16	2	22	4	9	2	9.5	3
9sym	9	1	–	8	3	62	5	8	3	13	4	6	3	5	5
alu2	10	8	–	70	4	22	3	45	5	102	7	38	5	19	5
alu4	14	8	1065	1244	7	413	6	159	12	–	–	162	7	323.5	14
apex4	9	19	931	1314	6	697	5	–	–	–	–	324	5	361.5	6
apex6	135	99	–	249	4	181	4	194	6	337	7	165	3	187.5	14
apex7	49	37	–	113	3	98	4	69	5	103	9	66	3	49.5	12
b9	16	5	–	55	3	36	3	43	3	57	4	35	3	38	9
clip	9	5	–	32	3	49	4	42	5	37	4	14	3	16	4
cout	35	16	–	50	3	50	3	34	5	52	5	36	3	36.5	9
duke2	22	29	–	208	3	177	4	180	7	268	6	158	4	188.5	12
ex1010	10	10	3567	4456	7	734	5	–	–	–	–	282	5	523	6
f51m	8	8	–	–	–	26	3	–	–	18	4	8	2	7	3
inc	7	9	–	–	–	27	3	–	–	51	4	14	2	19.5	2
misex1	8	7	–	14	2	16	2	14	2	19	3	9	2	8	2
misex3	14	14	980	1224	6	699	5	–	–	–	–	212	6	308	15
pdc	16	40	3222	4042	8	156	4	–	–	693	9	137	4	184	9
rd73	7	3	–	–	–	17	3	–	–	11	3	5	2	3.5	3
rd84	8	4	–	16	3	45	4	14	3	15	3	8	2	6.5	3
spla	16	46	2734	3243	8	237	5	–	–	398	9	199	5	264.5	11

(continued)

Table 15.2 (continued)

Benchmarks			DAOmap	DDBDD		ABC		BDS - PGA 2.0		FLDS		IRMA2FPGA		dekBDD	
Name	In	Out	Blocks	Blocks	Levels	Blocks	Levels	Blocks	Levels	Blocks	Levels	Blocks	Levels	Blocks	Levels
sqrt8	8	4	–	–	–	11	3	–	–	12	3	–	–	6.5	4
t481	16	1	–	5	2	13	4	5	2	5	2	5	2	4.5	5
Sum of blocks or Sum of levels			**12499**											**1964.5**	**61**
				16362	**77**									**2533**	**144**
						3789	**85**							**2569.5**	**156**
								823	**60**					**892**	**97**
										2213	**90**			**1053.5**	**115**
												1892	**73**	**2563**	**152**
Profit:			6.36	6.46	0.53	1.47	0.54	0.92	0.62	2.10	0.78	0.74	0.48	1.00	1.00

in Table 15.2, the results are also given for the systems: DAOmap [6] (the system directed at minimization of the logic levels using the methods for releasing the critical paths), DDBDD [7], ABC [8] (using the graphs AIG in the process of synthesis; it also enables resynthesis), and BDS–PGA 2.0 [9] (using BDD). The results obtained for the tool IRMA2FPGA [10, 11] were given for the blocks LUT45. In the bottom part of Table 15.2, the value of the total sum of the number of blocks and the number of levels for each system are given. Because the results are not given for all benchmarks, separate values of the total sums for the analyzed set of benchmarks (benchmarks that obtained from the results of two compared systems) are given for dekBDD to reliably compare the analyzed system with dekBDD. Knowing the values of the appropriate sums, the value of profit for a given system can be specified in accordance with Eq. (15.1) (see Footnote 1) [4].

$$profit = \frac{\text{the sum of the number of the blocks obtained for the system X}}{\text{the sum of the number of the blocks obtained for the dekBDD}} \quad (15.1)$$

When the value of the profit exceeds 1, the superiority of the system dekBDD over the compared system is demonstrated with respect of the number of blocks. Analogously, the profit is specified for the number of levels. The value of profit for both the number of blocks and the number of levels are presented in the last row in Table 15.2 (see Footnote 1) [4].

The values of the profits are presented in the form of the bar graph in Fig. 15.2 for the number of blocks (Fig. 15.2a) and the number of levels (Fig. 15.2b).

Comparing the system dekBDD with other academic systems with respect to the number of blocks, the proposed algorithm produces very good solutions. Compared with DAOmap or DDBDD, the results are approximately six times better. Only two systems (IRMA2FPGA, BDS–PGA 2.0) produce better solutions than dekBDD. Some systems are directed at the limitation of the number of levels, as shown in Fig. 15.2b, where the system dekBDD is poorly optimized with respect to the critical path.

Determining which cases dekBDD yields the best solutions and which cases are least efficient is difficult. The dekBDD system was created using the techniques known as the Decomp system for BDD. The Decomp system [12], which stems

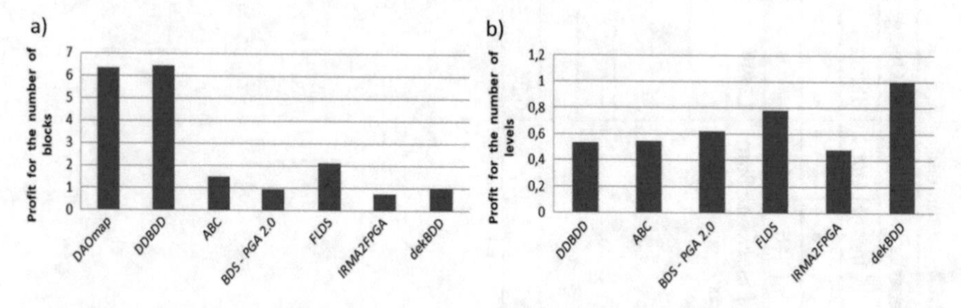

Fig. 15.2 Values of profits while comparing with the system dekBDD (LUT45) with respect to the number of blocks (**a**) and the number of levels (**b**) (see Footnote 1) [4]

from classic decomposition theory by Ashenhurst-Curtis, effectively coped with multioutput functions and incompletely specified functions. The results were worse in the case of difficult decomposable multioutput functions. The best results were obtained in the case of finding multiple decompositions that yielded better results considering both the number of LUTs and the number of levels. In the dekBDD system, the best results are obtained for symmetric functions (e.g., 9sym) or multi-output functions, in which symmetric functions occur (rd53, rd73, rd84). A drawback of the dekBDD system is related to incompletely specified functions. Unfortunately, a function description in the form of BDD does not enable easy searching for decompositions of these functions (see Footnote 1) [4].

Based on a synthetic comparison of the dekBDD system with other academic tools, the proposed method produces acceptable results with respect to logic blocks and relatively worse results with respect to the logic levels. A similar comparison of the MultiDec system with other academic systems is presented in [13–15].

The MultiDec system was compared with the popular ABC synthesis system. The ABC system is characterized by flexibility, and the results of the synthesis depend on the scripts, including the set of synthesis instructions. The MutliDec system is compared with three various scripts, such as ABC_1, ABC_2, ABC_3 and ABC_4. The scripts ABC_1 (strash; dch; if -K 5; mfs) and ABC_3 (strash; dch; if -K 6; mfs) enable technology mapping in logic blocks that have 5 inputs and 6 inputs, respectively. The script ABC_2 (strash; resyn2; fpga) enables the outcome of an advanced resynthesis and matching to logic blocks with 5 inputs. In addition, the experiment for the script ABC_4 (&st; &synch2; &if -K 5;) was conducted using the package ABC9 [15].

The experimental results are presented in the form of Table 15.3. For the MultiDec system in the first column (MultiDec_45), technology mapping was oriented to popular configurable blocks in the form of LUT4/2 or LUT5/1 (LUT_45). In the second column (MultiDec_56), technology mapping is directed to configurable blocks in the form of LUT5/2 or LUT6/1 (LUT_56) [15].

The results are compared based on the total number of blocks and levels included in the last row in Table 15.3. The values are presented in the form of bar charts, as shown in Fig. 15.3a (blocks) and Fig. 15.3b (levels) [15].

While comparing the obtained results regarding the number of blocks (Fig. 15.3a), the configurability of logic blocks is essential. Technology mapping that enables flexible selection of the number of inputs substantially reduces the number of necessary blocks compared with the solutions with a precisely defined number of inputs in a logic block. A comparison of the number of logic levels (Fig. 15.3b) indicates that the results differ from each other. A comparison of the total synthesis time indicates that the synthesis process in the ABC system is faster despite the lack of a resynthesis process in the MultiDec system [15].

Table 15.3 Comparison of MultiDec system with ABC system

Benchmarks			MultiDec_45			MultiDec_56			ABC_1			ABC_2			ABC_3			ABC_4		
Name	In	Out	LUT_45	Levels	T [ms]	LUT_56	Levels	T [ms]	LUT_5	Levels	T [ms]	LUT_5	Levels	T [ms]	LUT_6	Levels	T [ms]	LUT_5	Levels	T [ms]
5xp1	7	10	14	2	265	8	2	296	23	3	500	25	3	1110	15	2	480	18	2	390
b12	15	9	22.5	3	561	11.5	2	592	19	3	280	17	2	880	14	2	280	19	2	110
cm163a	16	5	6	3	670	5.5	2	452	9	2	220	9	2	830	7	2	220	9	2	160
cm85a	11	3	15	3	483	5	2	546	8	2	220	10	3	790	9	2	220	11	2	110
con1	7	2	2.5	2	46	1.5	1	15	3	2	170	3	2	780	2	1	170	3	2	130
f51m	8	8	15.5	4	249	5.5	2	156	26	3	260	34	3	910	13	3	270	24	3	140
inc	7	9	21	3	312	9	2	109	27	3	270	25	3	960	13	2	260	27	3	120
misex1	8	7	10.5	3	265	6.5	2	156	16	2	230	19	2	880	9	2	230	16	2	130
pcle	19	9	11.5	3	2340	9	3	3525	15	2	230	15	2	980	12	2	230	14	2	140
rd73	7	3	4.5	2	124	3	2	140	17	3	250	20	4	820	14	3	230	18	3	160
rd84	8	4	8	2	280	6	2	680	45	4	390	58	4	970	34	3	310	60	4	160
sqn	7	3	10.5	3	140	5	2	140	22	3	270	25	3	770	11	2	270	19	3	110
sqr6	6	11	12	2	140	6	1	31	21	3	310	24	3	970	11	1	250	20	2	140
sqrt8	8	4	6	2	124	4.5	2	78	11	3	280	12	3	840	9	3	230	12	3	110
t481	16	1	3.5	3	374	4.5	4	468	13	4	270	18	4	850	15	3	230	20	4	110
x2	10	7	12.5	3	280	6	2	680	13	2	250	14	2	860	12	2	200	14	2	130
z4m1	7	4	5.5	3	124	6	2	124	5	2	200	7	2	830	5	2	200	5	2	110
x5xp1	7	10	14	2	234	8	2	202	20	3	280	19	3	930	16	2	250	19	3	140
ldd	9	18	21	2	1185	18.5	2	1482	26	2	260	35	2	920	26	2	250	27	2	110
Sum:			216	50	8196	129	39	9872	339	51	5140	389	52	16880	247	41	4780	355	48	2710

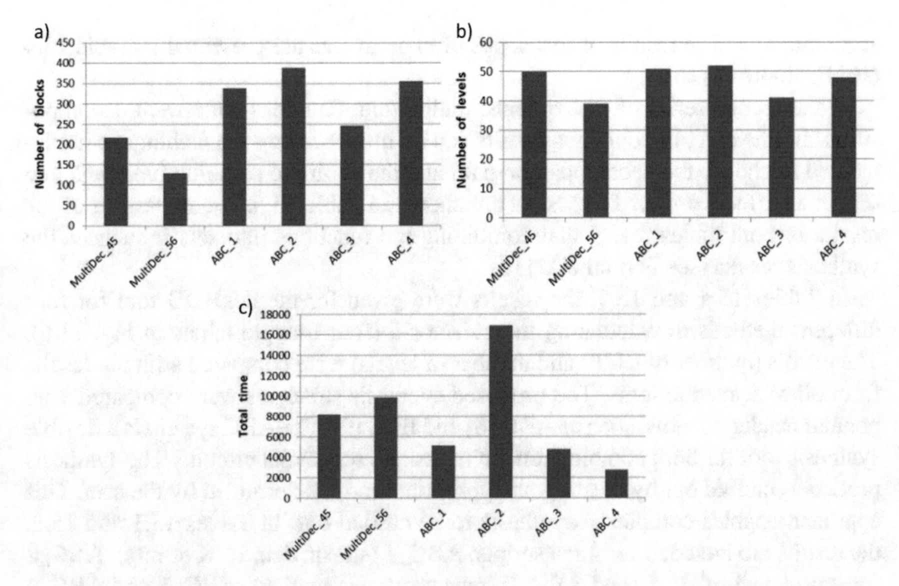

Fig. 15.3 Comparison of MultiDec system with ABC system for **a** total number of blocks, **b** total number of levels, and **c** total synthesis time

15.3 Effect of Triangle Tables on the Results

The dekBDD enabled us to carry out decomposition using the "*from input to output*" method, with a technology mapping into LUTs with a maximum of five inputs. The choice of an appropriate partition was carried out based on the analysis of the triangle tables presented in Fig. 11.6; The dekBDD enabled us to determine the required number of LUTs and the number of logic levels. The essence of the new synthesis ideas described in this book consists of two issues: searching for an appropriate decomposition path, and effective searching for shared resources for multioutput functions. Thus, two series of experiments were conducted[2] [16].

In the first series, experiments were only carried out for the set of single-output functions to compare these tools by taking into account only the efficiency of searching for an appropriate decomposition path rather than shared logic resources. Each basic benchmark that describes an m-input multioutput function was divided into m benchmarks, for which single-output functions were separately decomposed. The number of blocks was summed. The results pertained to a multioutput function, in which sharing of logic resources did not occur. These functions, as described in the form of .pla [1] or .blif files (BDS-PGA 2.0), were subjected to decomposition. Benchmarks were appropriately prepared before the process of decomposition. First, minimization [17] was carried out to describe single functions. Second, this description was performed to include each function in a separate file. This approach guaranteed that each function was separately decomposed in the case of each compared

[2]© Reprinted from Opara et al. [16], Copyright (2020), with permission from Elsevier.

academic tool. The results of this series of experiments are presented in Table 15.4 (see Footnote 2) [16].

In the second series of experiments, multioutput functions underwent decomposition. In this case, the obtained results can be influenced by the techniques implemented to choose the decomposition path and the ability to combine functions into sets in an effective way. The results are shown in Table 15.5. A comparison of the results in both tables reveal that combining the functions into sets influences the synthesis results (see Footnote 2) [16].

In Tables 15.4 and 15.5, the results were given for the dekBDD tool for four different methods of calculating the cofactor δ (four triangle tables in Fig. 11.6). The results (number of LUTs and number of levels) were compared with the results from other academic tools. The proposed synthesis strategies were compared with popular academic tools, such as ABC [8] and BDS [9]. The ABC system is a flexible synthesis tool for both combinational circuits and sequential circuits. The synthesis process is carried out by creating an appropriate script determined by the user. This approach enables complex resynthesis to be carried out. In Tables 15.1 and 15.2, the results are included for three scripts: ABC_1 (strash; dch; if -K 5; mfs), ABC_2 (resyn; resyn2; if -K 5) and ABC_3 (resyn; resyn2; if -K 6). ABC_1 and ABC_2 guarantee technology mapping to the blocks with five inputs. ABC_2 is particularly focused on the resynthesis process. In addition, the results for the LUT6 blocks are also presented (ABC_3) (see Footnote 2) [16].

In addition to an AIG net, the BDS–PGA 2.0 system also employs BDD to carry out decomposition. The system was launched with the default settings. The process of technology mapping itself was carried out using the Flow-Map tool (k = 5).

Experimental research was performed for the systems ABC, BDS–PGA 2.0. In addition, the results in Table 15.5 include some results from the literature for the DDBDD [18] and IRMA2FPGA [11] tools. DDBDD uses BDD for its description; IRMA2FPGA is a synthesis tool that effectively maps logic functions in FPGA (see Footnote 2) [16].

In both tables, the first three columns show the name of the benchmark, number of inputs and number of outputs. The following columns show the synthesis results for the various systems. The columns entitled 'Blocks' show the number of blocks, while the columns entitled 'Levels' show the number of logic levels in the critical path.

The last four columns—entitled dekBDD_x—provide the results using the proposed system with dekBDD for four different methods of calculating the cofactor δ. An objective comparison of academic systems is extremely difficult. An additional complication arises when the problem of mapping is considered in the blocks with a configurable number of inputs. Taking into account the reliability of comparison, we provide the following data for dekBDD each time: the number of LUT4 and LUT5 blocks, total number of blocks (LUT4 + LUT5) and number of blocks that are configurable, as shown in Spartan3 [19]. In these circuits, the blocks that are available may have one of two configurations: LUT5/1 and LUT4/2 (see Footnote 2) [16].

Table 15.4 Experimental results for the analysis of single-output functions (© Reprinted from Opara et al. [16], Copyright (2020), with permission from Elsevier)

Benchmarks			ABC_1 (strash; dch; if -K 5; mfs;)		ABC_2 (resyn; resyn2; if -K 5)		ABC_3 (resyn; resyn2; if -K 6)		BDS-pga 2.0		dekBDD_a (numb_of_g - card(Xb))				
Name	In	Out	Blocks	Levels	Blocks	Levels	Blocks	Levels	Blocks	Levels	LUT_4	LUT_5	LUT_4 + LUT_5	Blocks	Levels
5xp1	7	10	24	3	26	3	17	2	19	3	8	7	15	11	3
9sym	9	1	62	5	64	5	42	4	8	3	7	2	9	5.5	5
alu2	10	8	27	3	29	3	23	3	25	3	13	18	31	24.5	5
alu4	14	8	429	6	411	6	344	5	err		80	259	339	299	18
apex3	54	50	561	5	507	5	360	5	err		478	1385	1863	1624	36
apex7	49	37	182	4	122	4	99	4	err		57	109	166	137.5	16
b12	15	9	19	2	16	2	15	2	err		6	11	17	14	3
b9	16	5	53	3	38	3	35	2	err		21	37	58	47.5	7
bw	5	28	28	1	28	1	28	1	err		5	23	28	25.5	1
c8	28	18	35	2	34	3	28	3	err		12	26	38	32	5
cht	47	36	37	1	37	2	36	1	err		28	9	37	23	2
clip	9	5	44	3	40	3	31	3	27	3	6	13	19	16	4
cm162a	14	5	15	3	12	3	12	2	err		4	12	16	14	4
cm163a	16	5	9	2	9	2	7	2	err		3	7	10	8.5	3
cm85a	11	3	11	2	10	3	9	2	err		4	6	10	8	4
con1	7	2	3	2	3	2	2	1	3	2	1	2	3	2.5	2
count	35	16	60	3	51	3	37	3	err		8	66	74	70	8

(continued)

Table 15.4 (continued)

Benchmarks			ABC_1 (strash; dch; if -K 5; mfs;)		ABC_2 (resyn; resyn2; if -K 5)		ABC_3 (resyn; resyn2; if -K 6)		BDS-pga 2.0		dekBDD_a (numb_of_g - card(Xb))				
Name	In	Out	Blocks	Levels	Blocks	Levels	Blocks	Levels	Blocks	Levels	LUT_4	LUT_5	LUT_4 + LUT_5	Blocks	Levels
duke2	22	29	208	5	167	6	141	5	err		82	207	289	248	13
e64	65	64	651	3	258	4	231	3	err		48	496	544	520	16
ex1010	10	10	530	5	476	5	395	4	err		123	507	630	568.5	8
example2	85	66	171	3	108	3	100	3	err		67	120	187	153.5	6
f51m	8	8	24	3	33	3	19	3	13	3	7	5	12	8.5	3
inc	7	9	23	3	26	3	16	2	21	3	9	12	21	16.5	3
misex1	8	7	17	2	17	2	9	2	err		7	6	13	9.5	3
misex2	25	18	43	3	35	3	30	2	err		11	29	40	34.5	4
misex3	14	14	465	5	445	6	342	5	err		59	257	316	286.5	18
misex3c	14	14	163	5	161	6	125	5	err		54	169	223	196	17
mux	21	1	10	3	13	3	6	3	err		6	15	21	18	12
pcle	19	9	25	2	17	3	14	2	err		8	17	25	21	4
pdc	16	40	138	5	128	5	103	4	err		36	78	114	96	8
rd73	7	3	19	3	25	4	12	3	9	2	3	5	8	6.5	2
rd84	8	4	40	4	59	4	36	3	11	3	4	8	12	10	3
sao2	10	4	41	3	44	3	26	3	err		8	24	32	28	6

(continued)

Table 15.4 (continued)

Benchmarks			ABC_1 (strash; dch; if -K 5; mfs;)		ABC_2 (resyn; resyn2; if -K 5)		ABC_3 (resyn; resyn2; if -K 6)		BDS-pga 2.0		dekBDD_a (numb_of_g - card(Xb))				
Name	In	Out	Blocks	Levels	Blocks	Levels	Blocks	Levels	Blocks	Levels	LUT_4	LUT_5	LUT_4 + LUT_5	Blocks	Levels
sct	19	15	31	2	20	2	20	2	err		10	20	30	25	6
seq	41	35	884	6	669	6	559	5	err		1206	3979	5185	4582	54
spla	16	46	370	4	296	5	246	4	err		68	260	328	294	13
squar5	5	8	8	1	8	1	8	1	err		3	5	8	6.5	1
t481	16	1	13	4	20	4	21	3	5	2	1	4	5	4.5	5
table 3	14	14	703	6	559	6	461	5	err		194	614	808	711	17
table 5	17	15	746	6	578	6	494	5	err		425	1366	1791	1578.5	26
term1	34	10	60	4	41	4	30	4	err		15	49	64	56.5	9
ttt2	24	21	53	3	44	3	33	3	err		20	35	55	45	6
vda	17	39	425	4	244	5	251	4	err		92	374	466	420	14
vg2	25	8	76	4	52	4	43	4	err		22	49	71	60	11
x1	51	35	127	4	102	4	88	3	err		46	125	171	148	16
x2	10	7	14	2	14	2	12	2	17	3	7	7	14	10.5	4
x4	94	71	181	3	108	3	106	2	err		59	145	204	174.5	7
Z5xp1	7	10	22	1	26	3	18	2	err		8	7	15	11	3
z9sym	9	1	58	5	61	5	40	4	err		7	2	9	5.5	5

(continued)

Table 15.4 (continued)

Benchmarks			ABC_1 (strash; dch; if -K 5; mfs;)		ABC_2 (resyn; resyn2; if -K 5)		ABC_3 (resyn; resyn2; if -K 6)		BDS-pga 2.0		dekBDD_a (numb_of_g - card(Xb))				
Name	In	Out	Blocks	Levels	Blocks	Levels	Blocks	Levels	Blocks	Levels	LUT_4	LUT_5	LUT_4 + LUT_5	Blocks	Levels
Geometric mean:			62.07	3.07	55.50	3.37	43.56	2.80					56.67	46.85	6.27
Ratio versus ABC					**1.00**	**1.00**							**1.02**	**0.84**	**1.86**
Geom. mean (selected):									11.95	2.69			11.35	8.83	3.38
Ratio versus BDS-PGA2.0									**1.00**	**1.00**			**0.95**	**0.74**	**1.26**

Benchmarks			dekBDD_b (numb_of_g/card(Xb))					dekBDD_c (numb_of_bl - (card(Xb)numb_of_g))					dekBDD_d (2*numb_of_bl - card(Xb))				
Name	In	Out	LUT_4	LUT_5	LUT_4 + LUT_5	Blocks	Levels	LUT_4	LUT_5	LUT_4 + LUT_5	Blocks	Levels	LUT_4	LUT_5	LUT_4 + LUT_5	Blocks	Levels
5xp1	7	10	8	7	15	11	3	8	7	15	11	3	15	3	18	10.5	3
9sym	9	1	7	2	9	5.5	5	8	1	9	5	5	10	0	10	5	6
alu2	10	8	13	18	31	24.5	5	27	5	32	18.5	6	30	3	33	18	6
alu4	14	8	80	259	339	299	18	86	227	313	270	18	172	174	346	260	18
apex3	54	50	479	1384	1863	1623.5	35	589	1294	1883	1588.5	36	1112	1060	2172	1616	49
apex7	49	37	57	109	166	137.5	16	72	98	170	134	15	227	11	238	124.5	20
b12	15	9	6	11	17	14	3	6	11	17	14	3	22	3	25	14	5
b9	16	5	21	37	58	47.5	7	25	35	60	47.5	8	89	1	90	45.5	10

(continued)

Table 15.4 (continued)

Benchmarks			dekBDD_b (numb_of_g/card(Xb))					dekBDD_c (numb_of_bl - (card(Xb)numb_of_g))					dekBDD_d (2*numb_of_bl - card(Xb))				
Name	In	Out	LUT_4	LUT_5	LUT_4 + LUT_5	Blocks	Levels	LUT_4	LUT_5	LUT_4 + LUT_5	Blocks	Levels	LUT_4	LUT_5	LUT_4 + LUT_5	Blocks	Levels
bw	5	28	5	23	28	25.5	1	5	23	28	25.5	1	35	8	43	25.5	2
c8	28	18	12	26	38	32	5	12	26	38	32	5	54	0	54	27	9
cht	47	36	28	9	37	23	2	28	9	37	23	2	45	0	45	22.5	2
clip	9	5	6	13	19	16	4	9	11	20	15.5	5	20	6	26	16	5
cm162a	14	5	4	12	16	14	4	5	11	16	13.5	4	21	4	25	14.5	7
cm163a	16	5	3	7	10	8.5	3	3	7	10	8.5	3	13	1	14	7.5	4
cm85a	11	3	4	6	10	8	4	4	6	10	8	4	9	2	11	6.5	4
con1	7	2	1	2	3	2.5	2	1	2	3	2.5	2	5	0	5	2.5	2
count	35	16	8	66	74	70	8	8	66	74	70	8	133	1	134	67.5	15
duke2	22	29	82	207	289	248	13	157	121	278	199.5	13	233	78	311	194.5	12
e64	65	64	48	496	544	520	16	48	496	544	520	16	715	0	715	357.5	22
ex1010	10	10	125	521	646	583.5	8	127	503	630	566.5	8	236	449	685	567	8
example2	85	66	67	120	187	153.5	6	83	110	193	151.5	8	235	8	243	125.5	10
f51m	8	8	7	5	12	8.5	3	7	5	12	8.5	3	11	3	14	8.5	4
inc	7	9	9	12	21	16.5	3	12	9	21	15	3	15	7	22	14.5	3
misex1	8	7	7	6	13	9.5	3	7	6	13	9.5	3	10	4	14	9	3

(continued)

Table 15.4 (continued)

Benchmarks			dekBDD_b (numb_of_g/card(Xb))					dekBDD_c (numb_of_bl - (card(Xb)numb_of_g))					dekBDD_d (2*numb_of_bl - card(Xb))				
Name	In	Out	LUT_4	LUT_5	LUT_4 + LUT_5	Blocks	Levels	LUT_4	LUT_5	LUT_4 + LUT_5	Blocks	Levels	LUT_4	LUT_5	LUT_4 + LUT_5	Blocks	Levels
misex2	25	18	11	29	40	34.5	4	11	29	40	34.5	4	58	0	58	29	6
misex3	14	14	59	260	319	289.5	18	132	200	332	266	18	220	149	369	259	18
misex3c	14	14	54	169	223	196	17	84	142	226	184	18	143	111	254	182.5	18
mux	21	1	6	15	21	18	12	8	13	21	17	12	17	0	17	8.5	13
pcle	19	9	8	17	25	21	4	8	17	25	21	4	34	0	34	17	6
pdc	16	40	36	78	114	96	8	74	46	120	83	7	138	12	150	81	8
rd73	7	3	3	5	8	6.5	2	8	2	10	6	3	11	0	11	5.5	3
rd84	8	4	4	8	12	10	3	11	4	15	9.5	4	15	2	17	9.5	4
sao2	10	4	8	24	32	28	6	26	12	38	25	6	37	5	42	23.5	6
sct	19	15	10	20	30	25	6	10	20	30	25	6	40	5	45	25	7
seq	41	35	1251	3969	5220	4594.5	54	1467	3142	4609	3875.5	49	2634	2871	5505	4188	52
spla	16	46	68	260	328	294	13	102	214	316	265	10	312	95	407	251	12
squar5	5	8	3	5	8	6.5	1	3	5	8	6.5	1	9	2	11	6.5	2
t481	16	1	1	4	5	4.5	5	1	4	5	4.5	5	11	0	11	5.5	11
table 3	14	14	195	626	821	723.5	17	232	572	804	688	17	472	483	955	719	18
table 5	17	15	418	1378	1796	1587	26	479	1297	1776	1536.5	26	915	996	1911	1453.5	27

(continued)

Table 15.4 (continued)

Benchmarks			dekBDD_b (numb_of_g/card(Xb))					dekBDD_c (numb_of_bl - (card(Xb)numb_of_g))					dekBDD_d (2*numb_of_bl - card(Xb))				
Name	In	Out	LUT_4	LUT_5	LUT_4 + LUT_5	Blocks	Levels	LUT_4	LUT_5	LUT_4 + LUT_5	Blocks	Levels	LUT_4	LUT_5	LUT_4 + LUT_5	Blocks	Levels
term1	34	10	15	49	64	56.5	9	24	40	64	52	10	85	2	87	44.5	11
ttt2	24	21	20	35	55	45	6	21	34	55	44.5	6	65	6	71	38.5	9
vda	17	39	95	369	464	416.5	14	146	352	498	425	14	327	263	590	426.5	15
vg2	25	8	22	49	71	60	11	31	33	64	48.5	11	83	8	91	49.5	18
x1	51	35	46	125	171	148	16	73	97	170	133.5	15	236	13	249	131	17
x2	10	7	7	7	14	10.5	4	7	7	14	10.5	4	19	1	20	10.5	6
x4	94	71	59	145	204	174.5	7	66	139	205	172	7	291	4	295	149.5	11
Z5xp1	7	10	8	7	15	11	3	8	7	15	11	3	15	3	18	10.5	3
z9sym	9	1	7	2	9	5.5	5	8	1	9	5	5	10	0	10	5	6
Geometric mean:				56.73	46.90	6.27		57.35	44.66	6.41		72.94	42.09	7.90			
Ratio versus ABC				**1.02**	**0.85**	**1.86**		**1.03**	**0.80**	**1.90**		**1.31**	**0.76**	**2.34**			
Geom. mean(selected):				11.35	8.83	3.38		11.91	8.34	3.73		15.17	8.37	4.34			
Ratio versus BDS-PGA2.0				**0.95**	**0.74**	**1.26**		**1.00**	**0.70**	**1.39**		**1.27**	**0.70**	**1.62**			

Table 15.5 Experimental results for the analysis of multioutput functions (© Reprinted from Opara et al. [16], Copyright (2020), with permission from Elsevier)

Benchmarks			ABC_1 (strash; dch; if -K 5; mfs;)		ABC_2 (resyn; resyn2; if -K 5)		ABC_3 (resyn; resyn2; if -K 6)		BDS-pga 2.0		DDBDD (k5)		IRMA2FPGA (k5)		dekBDD_a (numb_of_g - card(Xb))				
Name	In	Out	Blocks	Levels	Blocks	Levels	Blocks	Levels	Blocks	Levels	Blocks	Levels	Blocks	Levels	LUT_4	LUT_5	LUT_4 + LUT_5	Blocks	Levels
5xp1	7	10	23	3	24	3	16	2	19	3	19	2	16	2	7	6	13	9.5	2
9sym	9	1	62	5	64	5	42	4	8	3	8	3	7	3	7	2	9	5.5	5
alu2	10	8	22	3	20	3	18	3	30	4	70	4	42	4	11	16	27	21.5	4
alu4	14	8	413	6	431	6	359	5	324	8	1244	7	161	7	103	309	412	360.5	18
apex3	54	50	528	5	552	5	399	5	err						786	2455	3241	2848	48
apex7	49	37	98	4	134	4	109	4	err		113	3	106	3	34	51	85	68	11
b12	15	9	19	3	16	2	14	2	18	3					7	12	19	15.5	4
b9	16	5	35	3	38	3	35	2	46	3	55	3	44	3	18	26	44	35	6
bw	5	28	28	1	28	1	28	1	28	1					5	23	28	25.5	1
c8	28	18	32	3	34	3	28	3	32	3					18	12	30	21	4
cht	47	36	37	2	37	2	36	1	37	2					28	9	37	23	2
clip	9	5	49	4	69	4	53	3	33	4	32	3	18	2	6	16	22	19	5
cm162a	14	5	11	3	12	3	12	2	err						3	8	11	9.5	4
cm163a	16	5	9	2	9	2	7	2	err						3	6	9	7.5	3
cm85a	11	3	8	2	10	3	9	2	err						3	5	8	6.5	3
con1	7	2	3	2	3	2	2	1	3	2					1	2	3	2.5	2
count	35	16	50	3	51	3	37	3	err		50	3	49	2	11	31	42	36.5	8
duke2	22	29	177	4	177	6	142	5	err		208	3	184	4	52	189	241	215	11
e64	65	64	221	4	258	4	231	3	168	11			305	3	48	54	102	78	16
ex1010	10	10	734	5	727	5	525	5	err		4456	7			122	511	633	572	8

(continued)

Table 15.5 (continued)

Benchmarks			ABC_1 (strash; dch; if -K 5; mfs;)		ABC_2 (resyn; resyn2; if -K 5)		ABC_3 (resyn; resyn2; if -K 6)		BDS-pga 2.0		DDBDD (k5)		IRMA2FPGA (k5)		dekBDD_a (numb_of_g - card(Xb))				
Name	In	Out	Blocks	Levels	Blocks	Levels	Blocks	Levels	Blocks	Levels	Blocks	Levels	Blocks	Levels	LUT_4	LUT_5	LUT_4 + LUT_5	Blocks	Levels
example2	85	66	101	3	107	3	99	3	err						53	69	122	95.5	6
f51m	8	8	18	3	25	3	16	3	13	3			14	2	6	4	10	7	3
inc	7	9	27	3	25	2	13	2	25	3					10	11	21	16	3
misex1	8	7	16	2	18	2	9	2	16	2	14	2	13	2	7	8	15	11.5	2
misex2	25	18	35	3	35	3	30	2	err				36	2	12	20	32	26	4
misex3	14	14	699	5	730	6	558	5	err		1224	6	236	6	58	323	381	352	14
misex3c	14	14	191	5	189	6	150	5	err				90	4	32	149	181	165	14
mux	21	1	10	3	13	3	6	3	11	3					6	15	21	18	12
pcle	19	9	16	3	17	3	14	2	21	3					5	13	18	15.5	4
pdc	16	40	156	4	172	4	141	3	294	7	4042	8			32	81	113	97	7
rd73	7	3	17	3	23	4	16	3	9	2			8	2	3	3	6	4.5	2
rd84	8	4	45	4	58	4	33	3	14	3	16	3	11	2	3	6	9	7.5	3
sao2	10	4	40	3	41	4	27	3	39	5			27	3	3	26	29	27.5	5
sct	19	15	20	2	23	3	23	2	22	2					10	10	20	15	4
seq	41	35	599	6	601	7	497	6	901	14	1518	6			2429	7541	9970	8755.5	66
spla	16	46	237	5	269	5	212	5	264	7	3243	8			50	234	284	259	9
squar5	5	8	8	1	8	1	8	1	8	1					3	5	8	6.5	1
t481	16	1	13	4	20	4	21	3	5	2	5	2			1	4	5	4.5	5
table3	14	14	566	6	587	6	486	5	err						208	574	782	678	17

(continued)

Table 15.5 (continued)

Benchmarks			ABC_1 (strash; dch; if -K 5; mfs;)		ABC_2 (resyn; resyn2; if -K 5)		ABC_3 (resyn; resyn2; if -K 6)		BDS-pga 2.0		DDBDD (k5)		IRMA2FPGA (k5)		dekBDD_a (numb_of_g - card(Xb))				
Name	In	Out	Blocks	Levels	Blocks	Levels	Blocks	Levels	Blocks	Levels	Blocks	Levels	Blocks	Levels	LUT_4	LUT_5	LUT_4 + LUT_5	Blocks	Levels
table5	17	15	546	6	527	7	438	6	err						402	1369	1771	1570	28
term1	34	10	29	4	38	4	28	4	53	5					11	26	37	31.5	9
ttt2	24	21	44	3	50	3	36	3	43	3					18	23	41	32	5
vda	17	39	314	4	309	5	234	4	393	6					192	554	746	650	12
vg2	25	8	44	4	52	4	43	4	68	5	80	4	44	3	4	38	42	40	10
x1	51	35	98	3	114	4	100	3	113	4					49	144	193	168.5	18
x2	10	7	13	2	14	2	13	2	16	3					5	9	14	11.5	3
x4	94	71	115	3	115	3	112	2	120	3					41	57	98	77.5	6
Z5xp1	7	10	20	3	20	3	16	2	err						7	6	13	9.5	2
z9sym	91	1	56	5	60	5	41	4	err						7	2	9	5.5	5
Geometric mean:			52.44	3.27	57.10	3.42	44.87	2.83									49.23	40.21	5.80
Ratio versus ABC			1.00	1.00													0.94	0.94	1.77
Geom. mean(selected):									36.38	3.38							35.45	28.69	4.95
Ratio versus BDS-PGA2.0									1.00	1.00							0.97	0.79	1.46
Geom. mean(selected):											136.38	3.83					71.48	59.91	7.23
Ratio versus DDBDD											1.00	1.00					0.52	0.44	1.89
Geom. mean(selected):													39.84	2.87			39.72	32.41	5.92

(continued)

Table 15.5 (continued)

Benchmarks			ABC_1 (strash; dch; if -K 5; mfs;)		ABC_2 (resyn; resyn2; if -K 5)		ABC_3 (resyn; resyn2; if -K 6)		BDS-pga 2.0		DDBDD (k5)		IRMA2FPGA (k5)		dekBDD_a (numb_of_g - card(Xb))				
Name	In	Out	Blocks	Levels	Blocks	Levels	Blocks	Levels	Blocks	Levels	Blocks	Levels	Blocks	Levels	LUT_4	LUT_5	LUT_4 + LUT_5	Blocks	Levels
Ratio versus IRMA2FPGA													1.00	1.00			1.00	0.81	2.06

Benchmarks			dekBDD_b (numb_of_g/card(Xb))					dekBDD_c (numb_of_bl - (card(Xb)numb_of_g))					dekBDD_d (2*numb_of_bl - card(Xb))				
Name	In	Out	LUT_4	LUT_5	LUT_4 + LUT_5	Blocks	Levels	LUT_4	LUT_5	LUT_4 + LUT_5	Blocks	Levels	LUT_4	LUT_5	LUT_4 + LUT_5	Blocks	Levels
5xp1	7	10	11	3	14	8.5	3	7	6	13	9.5	2	13	2	15	8.5	3
9sym	9	1	7	2	9	5.5	5	7	2	9	5.5	5	10	0	10	5	6
alu2	10	8	12	15	27	21	5	12	15	27	21	5	32	2	34	18	6
alu4	14	8	105	308	413	360.5	18	112	293	405	349	18	236	209	445	327	18
apex3	54	50	636	1785	2421	2103	40	784	2096	2880	2488	48	1198	1384	2582	1983	43
apex7	49	37	40	148	188	168	12	40	68	108	88	10	92	30	122	76	16
b12	15	9	6	11	17	14	3	7	10	17	13.5	3	22	2	24	13	4
b9	16	5	23	34	57	45.5	7	21	24	45	34.5	6	57	9	66	37.5	8
bw	5	28	5	23	28	25.5	1	5	23	28	25.5	1	5	23	28	25.5	1
c8	28	18	12	27	39	33	5	18	12	30	21	5	31	3	34	18.5	5
cht	47	36	28	9	37	23	2	28	9	37	23	2	45	0	45	22.5	2
clip	9	5	8	13	21	17	4	9	13	22	17.5	4	15	10	25	17.5	5
cm162a	14	5	3	12	15	13.5	4	10	2	12	7	4	10	2	12	7	4

(continued)

Table 15.5 (continued)

Benchmarks			dekBDD_b (numb_of_g/card(Xb))					dekBDD_c (numb_of_bl - (card(Xb)numb_of_g))					dekBDD_d (2*numb_of_bl - card(Xb))				
Name	In	Out	LUT_4	LUT_5	LUT_4 + LUT_5	Blocks	Levels	LUT_4	LUT_5	LUT_4 + LUT_5	Blocks	Levels	LUT_4	LUT_5	LUT_4 + LUT_5	Blocks	Levels
cm163a	16	5	2	9	11	10	3	3	6	9	7.5	3	10	2	12	7	4
cm85a	11	3	4	6	10	8	4	3	5	8	6.5	3	8	2	10	6	4
con1	7	2	1	2	3	2.5	2	1	2	3	2.5	2	5	0	5	2.5	2
count	35	16	11	56	67	61.5	6	9	29	38	33.5	9	53	4	57	30.5	11
duke2	22	29	79	205	284	244.5	14	48	170	218	194	11	238	79	317	198	14
e64	65	64	83	24	107	65.5	17	50	52	102	77	16	52	51	103	77	17
ex1010	10	10	124	527	651	589	8	122	511	633	572	8	237	452	689	570.5	8
example2	85	66	68	125	193	159	7	57	64	121	92.5	5	126	21	147	84	9
f51m	8	8	9	2	11	6.5	3	6	4	10	7	3	10	2	12	7	4
inc	7	9	7	12	19	15.5	2	11	9	20	14.5	3	16	4	20	12	3
misex1	8	7	5	6	11	8.5	2	10	3	13	8	2	10	3	13	8	2
misex2	25	18	12	20	32	26	4	18	18	36	27	4	38	4	42	23	6
misex3	14	14	55	323	378	350.5	14	77	298	375	336.5	14	250	176	426	301	14
misex3c	14	14	34	153	187	170	13	33	147	180	163.5	14	88	109	197	153	14
mux	21	1	6	15	21	18	12	6	15	21	18	12	17	0	17	8.5	13
pcle	19	9	7	17	24	20.5	4	7	11	18	14.5	4	18	4	22	13	5
pdc	16	40	39	75	114	94.5	8	32	73	105	89	6	139	8	147	77.5	7
rd73	7	3	5	1	6	3.5	3	3	3	6	4.5	2	5	1	6	3.5	3
rd84	8	4	7	5	12	8.5	4	3	6	9	7.5	3	10	1	11	6	3

(continued)

Table 15.5 (continued)

Benchmarks			dekBDD_b (numb_of_g/card(Xb))					dekBDD_c (numb_of_bl - (card(Xb)numb_of_g))					dekBDD_d (2*numb_of_bl - card(Xb))				
Name	In	Out	LUT_4	LUT_5	LUT_4 + LUT_5	Blocks	Levels	LUT_4	LUT_5	LUT_4 + LUT_5	Blocks	Levels	LUT_4	LUT_5	LUT_4 + LUT_5	Blocks	Levels
sao2	10	4	9	24	33	28.5	6	8	26	34	30	6	23	17	40	28.5	6
sct	19	15	16	9	25	17	6	10	10	20	15	4	21	5	26	15.5	7
seq	41	35	1818	5601	7419	6510	57	1251	3963	5214	4588.5	48	2558	3167	5725	4446	56
spla	16	46	91	202	293	247.5	10	69	214	283	248.5	11	229	109	338	223.5	10
squar5	5	8	3	5	8	6.5	1	3	5	8	6.5	1	3	5	8	6.5	1
t481	16	1	1	4	5	4.5	5	1	4	5	4.5	5	11	0	11	5.5	11
table3	14	14	213	582	795	688.5	17	217	574	791	682.5	17	494	392	886	639	18
table5	17	15	403	1373	1776	1574.5	27	416	1348	1764	1556	28	1020	1008	2028	1518	29
term1	34	10	18	32	50	41	9	15	15	30	22.5	7	29	9	38	23.5	10
ttt2	24	21	26	18	44	31	5	19	19	38	28.5	5	38	8	46	27	5
vda	17	39	132	431	563	497	14	165	492	657	574.5	14	370	350	720	535	15
vg2	25	8	13	45	58	51.5	12	15	59	74	66.5	13	75	22	97	59.5	20
x1	51	35	41	114	155	134.5	14	43	111	154	132.5	14	176	29	205	117	17
x2	10	7	10	4	14	9	4	10	4	14	9	4	11	3	14	8.5	4
x4	94	71	61	139	200	169.5	7	47	57	104	80.5	7	118	14	132	73	9
Z5xp1	7	10	11	3	14	8.5	3	7	6	13	9.5	2	13	2	15	8.5	3
z9sym	91	1	7	2	9	5.5	5	7	2	9	5.5	5	10	0	10	5	6
Geometric mean:					53.67	42.60	6.12			48.46	38.25	5.80			57.33	35.28	6.91
Ratio versus ABC					1.02	0.81	1.87			0.92	0.73	1.78			1.09	0.67	2.11

(continued)

Table 15.5 (continued)

Benchmarks			dekBDD_b (numb_of_g/card(Xb))					dekBDD_c (numb_of_bl - (card(Xb)numb_of_g))					dekBDD_d (2*numb_of_bl - card(Xb))				
Name	In	Out	LUT_4	LUT_5	LUT_4 + LUT_5	Blocks	Levels	LUT_4	LUT_5	LUT_4 + LUT_5	Blocks	Levels	LUT_4	LUT_5	LUT_4 + LUT_5	Blocks	Levels
Geom. mean(selected):					37.38	28.91	5.27			34.52	26.97	4.98			41.12	24.92	5.88
Ratio versus BDS-PGA2.0					1.03	0.79	1.56			0.95	0.74	1.47			1.13	0.68	1.74
Geom. mean(selected):					78.90	64.41	7.71			70.40	57.44	7.24			88.24	54.08	8.85
Ratio versus DDBDD					0.58	0.47	2.01			0.52	0.42	1.89			0.65	0.40	2.31
Geom. mean(selected):					44.86	34.54	6.46			41.26	32.62	6.07			48.79	30.08	7.40
Ratio versus IRMA2FPGA					1.13	0.87	2.25			1.04	0.82	2.11			1.22	0.76	2.58

At the bottom of Tables 15.4 and 15.5, values for the geometric mean of the appropriate parameters and cofactors (ratio vs. ABC and ratio vs. BDS-PGA2.0) are listed. These values enable a comparison of each system with the proposed synthesis algorithms implemented in dekBDD. Values lower than one indicate the superiority of dekBDD (see Footnote 2) [16].

In the case of BDS–PGA 2.0, an unexpected problem arose. In the case of the benchmarks that describe single functions, the process of synthesis could not be carried out due to the following error: "free(): invalid next, malloc(): memory corruption or core dump." In this situation, comparing BDS–PGA 2.0 with other systems is not sufficiently reliable (see Footnote 2) [16].

In addition to the direct comparison shown in Tables 15.4 and 15.5, the results are presented in the form of bar graphs (Figs. 15.4, 15.5, 15.6 and 15.7). The geometric average was presented for the number of blocks and levels. When comparing the

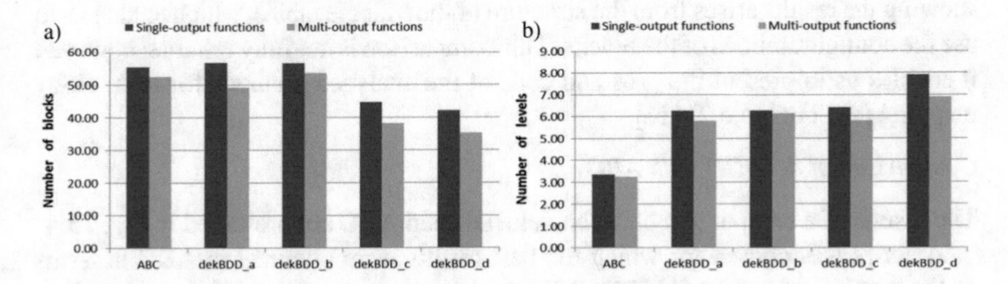

Fig. 15.4 Comparison of the results obtained with ABC (see Footnote 2) [16]

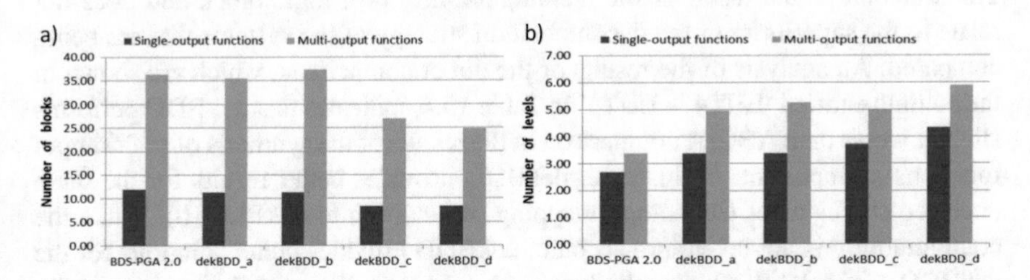

Fig. 15.5 Comparison of the results obtained with BDS-PGA 2.0 (see Footnote 2) [16]

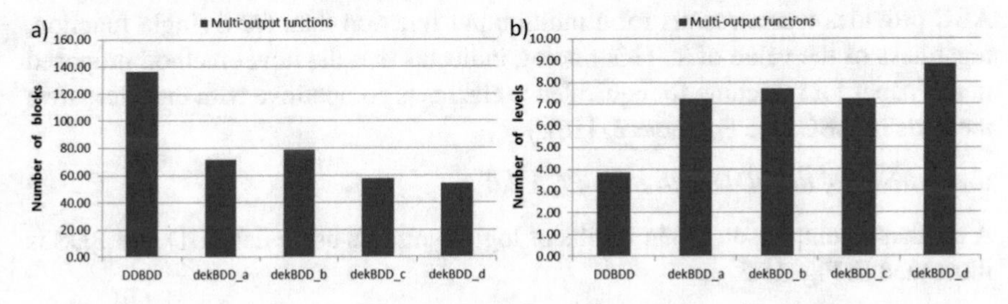

Fig. 15.6 Comparison of the results obtained with DDBDD (see Footnote 2) [16]

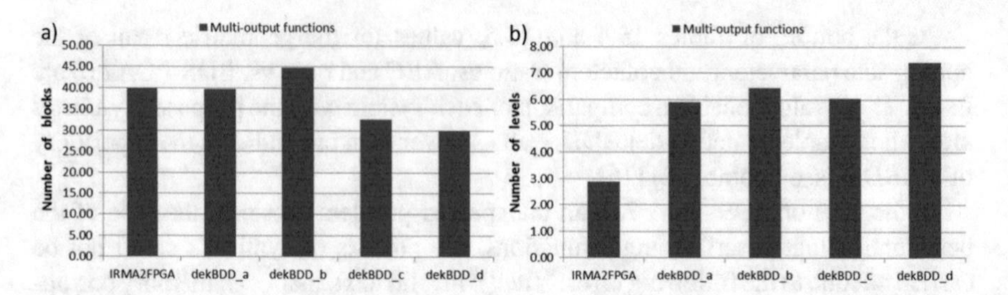

Fig. 15.7 Comparison of the results obtained with IRMA2FPGA (see Footnote 2) [16]

number of blocks for dekBDD_a and dekBDD_b, the values in the diagrams refer to the number of LUT4 + LUT5 blocks. In the case of dekBDD_c and dekBDD_d, the size of the columns shows the number of blocks for Spartan3 [19]. This method of showing the results arises from the structure of the triangle tables, which enable us to use the configurabilities of the blocks. This comparison is not fully reliable; however, it enables us to present the pros and cons of the analyzed methods for technology mapping (see Footnote 2) [16].

Comparison of dekBDD with ABC

The results of a comparison between dekBDD and ABC are presented in Fig. 15.4

A script was chosen for which the best results were obtained in ABC in terms of the number of blocks (LUT5). When conducting a synthesis of a single-output function, dekBDD provides better results for dekBDD_c and dekBDD_d (Fig. 15.4a). This outcome is the result of the reconfigurability of a logic block and does not relate to the superiority of the decomposition strategy of the systems that are being compared. An analysis of the results of the direct comparison, which are shown in the column entitled 'LUT4 + LUT5' in Table 15.4, indicates that dekBDD performs slightly worse than ABC. A comparison of the results of the synthesis of multioutput functions is important. In this case, dekBDD provides better results for the three analyzed strategies of technology mapping (in addition to dekBDD_b). When the configurability is not taken into account, dekBDD provides better solutions for the dekBDD_a and dekBDD_c methods (LUT4 + LUT5). When analyzing Fig. 15.4b, dekBDD performs substantially worse for each case in terms of the number of logic levels. For scripts ABC_2 and ABC_3 (where resynthesis is especially employed), ABC provides worse results for a multioutput function than for a single function, regardless of the value of k. This finding indicates that the novel method proposed in this paper for searching for equivalence classes is competitive with the alternative methods in ABC (see Footnote 2) [16].

Comparison of dekBDD with BDS-PGA 2.0

A synthetic comparison of the results of logic synthesis using dekBDD and BDS is illustrated in Fig. 15.5.

The results are almost equivalent for the single-output function. The better outcome for dekBDD_c and dekBDD_d is obtained by considering the configurabilities of blocks. When analyzing only the number of blocks (LUT4 + LUT5), dekBDD_d provides worse solutions than dekBDD_a and dekBDD_b. For multi-output functions, dekBDD_a and dekBDD_c always yield better results. When comparing the number of logic levels (Fig. 15.5b), BDS-PGA 2.0 is substantially better (see Footnote 2) [16].

The dekBDD system is very effective for multioutput functions in terms of the number of logic blocks and yields substantially better solutions than BDS-PGA 2.0.

Comparison of dekBDD with DDBDD

A comparison of the results obtained from implementing benchmarks in dekBDD with the results obtained for DDBDD are presented in Fig. 15.6 (see Footnote 2) [16].

In the case of multioutput functions, dekBDD provides much better solutions in terms of the number of blocks than DDBDD for all methods of technology mapping. A lower number of blocks generates a higher number of logic levels (Fig. 15.6b). The obtained results are not surprising since DDBDD is directed at minimizing the number of levels (see Footnote 2) [16, 18].

Comparison of dekBDD with IRMA2FPGA

A comparison of dekBDD with IRMA2FPGA is illustrated in Fig. 15.7.

When comparing the results obtained for dekBDD and IRMA2FPGA in terms of the number of blocks, almost no difference can be observed. IRMA2FPGA is slightly better than dekBDD_b; in the case of dekBDD_c and dekBDD_d, the method of assessing results depends on the way the blocks are counted. IRMA2FPGA provides substantially better results than dekBDD in terms of the number of logic levels (see Footnote 2) [16].

Comparison of tools in terms of synthesis time

A key element of assessing the usefulness of a synthesis tool is synthesis time. A series of experiments was conducted with the aim of comparing synthesis times for the same synthesis circuits using various tools. These experiments were conducted on the same computer with an Intel Core i5-3210 M 2.5 GHz processor and 8 GB RAM memory (see Footnote 2) [16].

In these experiments, special attention was paid to the benchmarks with the highest number of inputs shown in Tables 15.4 and 15.5. The synthesis times for eight benchmarks are expressed in seconds and are presented in Table 15.6 (see Footnote 2) [16].

The results are shown for ABC (two scripts for which $k = 5$), BDS-PGA 2.0 and dekBDD. A reliable comparison with other tools (DDBDD, IRMA2FPGA) was impossible, as the results presented in Table 15.2 were obtain from the literature (see Footnote 2) [16].

Based on the results shown in Table 15.6, ABC synthesizes circuits substantially faster than dekBDD. A comparison of the analyzed methods with BDS-PGA 2.0

Table 15.6 Comparison of the analyzed algorithms in terms of synthesis time (see Footnote 2) [16]

Benchmarks			ABC_1 (strash; dch; if -K 5; mfs;)	ABC_2 (resyn; resyn2; if -K 5;)	BDS-PGA 2.0	dekBDD
Name	In	Out				
apex3	54	50	1.59	1.64	err	12.68
apex7	49	37	0.41	1.17	err	0.72
cht	47	36	0.33	1.05	0.21	0.01
e64	65	64	0.72	1.17	1.24	3.04
example2	85	66	0.38	1.15	err	0.81
seq	41	35	1.87	1.61	6.2	23
x1	51	35	0.37	1.18	0.37	0.43
x4	94	71	0.41	1.7	0.42	0.59
Sum:			6.08	10.67	8.44	41.28

is more difficult. Although BDS-PGA 2.0 is slightly more effective, determining this finding is difficult since synthesis was a failure for three of the eight cases (see Footnote 2) [16].

The following points sum the results of the comparison:

- dekBDD yields efficient solutions in terms of the number of blocks compared with other tools. The multioutput function is the most advantageous, which indicates a high efficiency for the analyzed algorithm in searching for equivalence classes.
- The results of the synthesis in dekBDD entirely depends on the method of technology mapping, which is precisely determined by the structure of a triangle table. The proposed strategy of technology mapping enables us to match the synthesis algorithm to its technological limitations.
- The main drawback of dekBDD is a weak optimization of the obtained solutions in terms of the number of logic levels. This problem may be solved by releasing the critical path in the process of resynthesis. Currently, assessing the influence of resynthesis on the results obtained using dekBDD is difficult. Resynthesis may worsen the results in terms of the number of blocks and may prolong the synthesis time.

Taking into account the pros and cons of dekBDD, using the algorithms analyzed in this paper in the process of optimizing modules of a circuit designed with a limited number of inputs is better (see Footnote 2) [16].

15.4 Comparison of the MultiDec System with Commercial Systems

The comparison of MultiDec system with commercial tools is substantially more precise. The MultiDec system enables the creation of a structural description of the results in HDL and enables the last stages of synthesis in commercial tools. A series of experiments were performed using commercial tools to compare the effects of synthesis for the cases in which decomposition was performed in the MultiDec system with the solutions, in which the total synthesis process was conducted using a commercial tool. The results were compared with the most popular and widely available synthesis tools, such as Quartus 15.1 (Altera) and ISE 14.7 (Xilinx). In the case of Quartus, synthesis was oriented to the ALM blocks included in the Stratix and Cyclon series V devices. In the case of ISE, synthesis was directed to the blocks included in the Artix, Kintex series 7, and Virtex series 6 devices (are characterized by the same configurable abilities). The results of the comparison are presented in Table 15.7 [15].

In the second table, the first two columns describe the synthesis results (number of blocks) for ISE. The column 'ISE' describes the case in which the total synthesis process was performed in the ISE system. The column ABC_3 + ISE presents the results of the synthesis gained using the system ABC and ISE tool. The column 'MultiDec' + 'ISE' includes the solutions in which an HDL circuit was subjected to synthesis and was gained as the result of decomposition performed using MultiDec. In the next two columns, the results gained using Quartus are presented. The last row of the second table determines the total number of blocks for the compiled set of benchmarks obtained in each case [15].

While comparing the total number of blocks gained in both cases for ISE, the results are nearly identical. Thus, few advantages are achieved by the MultiDec algorithm. In the Quartus system, reducing the number of blocks to approximately 17% by the decomposition methods included in the MultiDec system was possible. The elements of matching decomposition algorithms to logic resources of FPGA structure included in the MultiDec system improved the efficiency of the process of synthesis [15].

15.5 Synthesis of Sequential Circuits

In a similar way, experiments can be performed for the combinational part of sequential circuits. The synthesis in the Quartus tool includes descriptions of excitation blocks and output blocks; the results are summarized in Tables 15.8 and 15.9, respectively. Synthesized descriptions were generated in the MultiDec or ABC system.

Table 15.7 Comparison of the synthesis results regarding the number of blocks for commercial tools

Benchmarks			ISE	ABC_3 + ISE	MultiDec + ISE	Quartus	ABC_3 + Quartus	MultiDec + Quarus
Name	In	Out						
5xp1	7	10	14	14	13	11	9	8
b12	15	9	16	15	20	10	9	13
cm163a	16	5	7	7	10	5	5	6
cm85a	11	3	6	6	8	5	6	10
con1	7	2	2	2	2	1	1	1
f51m	8	8	13	14	10	8	9	6
inc	7	9	11	11	13	11	11	11
misex1	8	7	9	9	9	6	6	7
pcle	19	9	12	12	12	9	9	8
rd73	7	3	19	11	6	8	6	3
rd84	8	4	13	14	9	41	26	6
sqn	7	3	6	6	9	8	8	8
sqr6	6	11	10	10	10	7	7	7
sqrt8	8	4	11	11	6	6	8	4
t481	16	1	4	10	8	3	3	6
x2	10	7	11	11	11	7	8	10
z4ml	7	4	5	5	11	5	3	5
x5xp1	7	10	14	15	16	11	10	10
ldd	9	18	26	24	27	15	14	18
Sum:			209	207	210	177	158	147

When comparing both systems and considering the number of logic blocks needed to implement separate benchmarks, the number of blocks after decomposition ('Blocks') is lower than that obtained after the last synthesis stages in Quartus ('Quartus'). In this case, the results obtained for MultiDec are slightly better but are substantially worse than the predictions obtained in MultiDec. The results of the synthesis indicate the substantial advantages of using the configurabilities of blocks that effectively utilize MultiDec. The obtained results are almost the same when the number of logic levels is taken into account. When the decomposition times are compared, MultiDec substantially improves [20].

Table 15.8 Comparison of MultiDec with ABC (δ function)

			delta							
Benchmarks			MultiDec				ABC			
Name	In	Out	Blocks	Quartus	Levels	Time [ms]	Blocks	Quartus	Levels	Time [ms]
Beecount	6	3	3	3	1	15	3	3	1	260
dk14	6	3	3	2	1	15	3	3	1	230
dk15	5	2	1	1	1	15	2	1	1	160
dk16	7	5	12	12	2	265	18	13	2	200
dk17	10	8	12	7	2	1466	19	11	2	160
dk27	4	3	2	2	1	15	3	2	1	170
dk512	5	4	2	2	1	15	4	2	1	140
Donfile	7	5	6	6	2	62	7	6	2	160
ex2	7	5	8	11	2	202	14	12	2	190
ex3	6	4	4	4	1	15	4	4	1	140
ex4	10	4	6	9	2	312	9	7	2	140
ex5	6	4	4	4	1	31	4	4	1	140
ex7	6	4	4	4	1	31	4	4	1	140
Lion	4	2	1	1	1	15	2	1	1	120
lion9	6	4	4	4	1	15	4	4	1	140
Mc	5	2	1	1	1	0	2	1	1	130
s8	7	3	4	5	2	140	8	7	2	130
s27	7	3	3	3	1	15	3	3	1	140
Shiftreg	4	3	2	2	1	15	3	2	1	110
Tav	6	2	1	1	1	15	1	1	1	140
train4	4	2	1	1	1	15	2	1	1	120
train11	6	4	4	4	1	31	4	4	1	130
Sum:			88	89	28	2720	123	96	28	3390

15.6 Technology Mapping in Complex Logic Blocks—Results of Experiments

Experiments focused on technological mapping were performed in complex logic blocks for the example of ALM blocks. The synthesis was conducted using a commercial tool named Quartus II. A set of benchmarks was synthesized and described in the form of equations in Verilog HDL. Descriptions were generated from .pla (withoutMD) or from the descriptions .pla, which were initially decomposed based on the methods of technology mapping in the tool MultiDec (withMD) [20].

In the first series of experiments, combinational circuits were analyzed; the results are shown in Table 15.10. Table 15.10 includes the number of ALM blocks and the

Table 15.9 Comparison of MultiDec with ABC (λ function)

Benchmarks			lambda							
			MultiDec				ABC			
Name	In	Out	Blocks	Quartus	Levels	Time [ms]	Blocks	Quartus	Levels	Time [ms]
Beecount	6	4	4	4	1	46	4	3	1	220
dk14	6	5	5	4	1	46	5	5	1	170
dk15	5	5	3	3	1	31	5	3	1	110
dk16	7	3	5	3	2	78	6	5	2	140
dk17	10	3	5	4	2	280	10	7	3	130
dk27	4	2	1	1	1	15	2	1	1	160
dk512	5	3	2	2	1	15	3	2	1	190
Donfile	7	1	1	1	1	15	1	1	1	80
ex2	7	2	2	2	2	78	3	3	2	140
ex3	6	2	2	2	1	15	2	2	1	130
ex4	10	9	7	7	1	21	9	8	1	140
ex5	6	2	2	1	1	31	1	1	1	140
ex7	6	2	2	1	1	15	1	1	1	130
lion	4	1	1	1	1	15	1	1	1	110
lion9	6	1	1	1	1	15	1	1	1	140
Mc	5	5	1	2	1	15	3	2	1	120
s8	7	1	1	1	2	46	2	1	2	140
s27	7	1	1	2	1	15	1	1	1	140
Shiftreg	4	1	1	1	1	15	1	1	1	120
Tav	6	4	4	2	1	15	4	2	1	120
train4	4	1	1	1	1	15	1	1	1	120
train11	6	1	1	1	1	15	1	1	1	120
Sum:			53	47	26	852	67	53	27	3010

number of particular LUT (ALUT) blocks that have a given number of inputs included in the ALM blocks. In addition, the number of logic levels (depth) is given. The last row shows the sum of the number of levels and the number of separate blocks. The total number of blocks is presented in the form of a graph in Fig. 15.8 [20].

An analysis of the graph in Fig. 15.8 indicates that the proposed methods of technology mapping caused a substantial reduction in the number of ALM blocks. The number of the various integral LUT blocks was reduced (apart from LUT5), and the greatest reduction was obtained for LUTs with six and four inputs. As a result of using the proposed methods, the number of logic levels increased by approximately 10%, which is unfavorable from the point of view of the dynamic behavior of a circuit [20].

Table 15.10 Results of synthesis for combinational circuits

Benchmarks			without MD							with MD						
Name	In	Out	ALM	ALUT7	ALUT6	ALUT5	ALUT4	ALUT $\Leftarrow$ 3	Levels	ALM	ALUT7	ALUT6	ALUT5	ALUT4	ALUT $\Leftarrow$ 3	Levels
5xp1	7	10	11	0	5	3	2	6	2	9	0	4	1	1	7	2
alu2	10	8	15	0	5	11	5	4	2	13	1	4	6	3	6	3
b12	15	9	10	1	4	3	3	4	2	12	0	5	8	1	5	3
cm163a	16	5	5	1	2	2	1	0	2	7	1	3	3	1	2	3
f51m	8	8	12	0	5	4	5	5	3	6	0	2	3	2	3	2
inc	7	9	11	0	8	2	1	2	2	11	0	8	2	1	3	3
ldd	9	19	17	0	4	4	10	11	4	19	0	8	2	7	13	2
misex1	8	7	6	2	3	0	2	0	1	7	0	4	2	4	0	2
misex2	25	18	18	1	6	9	8	4	3	20	0	12	8	3	5	3
pcle	19	9	9	1	5	2	1	2	2	9	2	4	2	1	2	3
rd73	7	3	8	0	6	0	0	3	2	3	0	0	3	2	1	2
rd84	8	4	52	2	32	11	12	12	4	7	0	4	1	1	3	3
sct	19	15	12	0	3	5	7	5	2	14	1	3	9	5	6	3
sqn	7	3	8	0	6	0	0	3	2	8	0	6	0	0	3	2
sqr6	6	11	7	0	4	3	1	2	1	8	0	4	3	1	3	2
sqrt8	8	4	2	0	2	0	1	4	3	2	0	4	1	1	2	2
t481	16	1	3	0	0	0	5	1	3	6	0	2	5	2	0	4
x2	10	7	8	0	2	3	2	6	2	10	0	3	6	4	4	3
Sum:			214	8	102	62	66	74	42	171	5	80	65	40	68	47

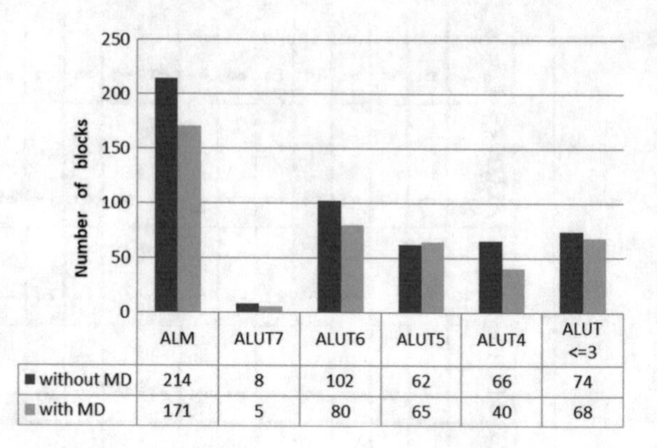

Fig. 15.8 Number of logic blocks obtained after the synthesis of combinational circuits

In the second series of experiments, combinational blocks of sequential circuits were synthesized, and the blocks δ and λ were separately synthesized. Coding of inner states and the use of a natural binary code were assumed. The results obtained for the δ blocks are shown in Table 15.11, and those for the λ blocks are shown in Table 15.12 [20].

The results of the synthesis of sequential circuits are presented in synthetic form in the graphs in Figs. 15.9 and 15.10.

From the point of view of the experiments, the key problem in the synthesis of combinational blocks of FSMs is that the circuits are so small that they do not need decomposition. In the case of blocks with a higher number of inputs, using decomposition with the proposed methods of technology mapping enables (for blocks δ and λ) a reduction in the number of required ALM blocks. In both cases, the largest reduction is obtained for LUTs with six inputs. An analysis of the obtained number of logic levels shows a rapid increase in the number of levels when the proposed methods are employed, in the case of δ blocks. For the λ block, the number of levels remains the same [20].

The most important indicator that confirms the advantages of the proposed method is the fact that the total number of ALM blocks is lower when using the proposed method of technology mapping. The situation is the same for both combinational circuits and sequential circuits. Thus, a key element in the technology mapping of the circuits implemented in FPGA is the ability to use available configurations of ALMs [20].

A greater number of results obtained is shown in [3, 4, 12–16, 20, 21].

Table 15.11 Results of the synthesis of transition blocks (δ function)

Benchmarks			Delta													
			Without MD							With MD						
Name	In	Out	ALM	ALUT7	ALUT6	ALUT5	ALUT4	ALUT ⇐ 3	Levels	ALM	ALUT7	ALUT6	ALUT5	ALUT4	ALUT ⇐ 3	Levels
beecount	6	3	3	0	3	0	0	0	1	3	0	3	0	0	0	1
dk14	6	3	3	0	3	0	0	0	1	2	0	0	0	2	1	1
dk15	5	2	1	0	0	2	0	0	1	1	0	0	0	2	0	1
dk16	7	5	13	0	10	0	0	5	2	12	0	9	1	0	5	2
dk17	10	8	11	0	5	3	6	3	2	7	0	3	2	5	1	2
dk27	4	3	2	0	0	0	3	0	1	2	0	0	0	2	1	1
dk512	5	4	2	0	0	4	0	0	1	2	0	0	4	0	0	1
donfile	7	5	2	0	4	0	2	2	2	6	0	4	0	2	1	2
ex2	7	5	12	0	9	0	0	5	2	11	0	6	4	1	4	3
ex3	6	4	4	0	4	0	0	0	1	4	0	4	0	0	0	1
ex4	10	4	7	3	2	0	2	1	2	9	0	5	5	1	2	3
ex5	6	4	4	0	4	0	0	0	1	4	0	4	0	0	0	1
ex7	6	4	4	0	4	0	0	0	1	4	0	3	1	0	0	1
lion	4	2	1	0	0	0	2	0	1	1	0	0	0	1	1	1
lion9	6	4	4	0	4	0	0	0	1	4	0	4	0	0	0	1
mc	5	2	1	0	0	1	0	1	1	1	0	0	1	0	1	1
s8	7	3	7	0	4	2	1	2	2	5	0	2	5	0	1	2
s27	7	3	3	0	2	1	0	0	1	3	0	2	1	0	1	2
shiftreg	4	3	2	0	0	0	2	1	1	2	0	0	0	2	1	1

(continued)

Table 15.11 (continued)

Benchmarks			Delta													
			Without MD							With MD						
Name	In	Out	ALM	ALUT7	ALUT6	ALUT5	ALUT4	ALUT ⇐ 3	Levels	ALM	ALUT7	ALUT6	ALUT5	ALUT4	ALUT ⇐ 3	Levels
tav	6	2	1	0	0	0	0	2	1	1	0	0	0	0	2	1
train4	4	2	1	0	0	0	1	1	1	1	0	0	0	1	1	1
train11	6	4	4	0	4	0	0	0	1	4	0	4	0	0	0	1
Sum:			92	3	62	13	19	23	28	89	0	53	24	19	23	31

Table 15.12 Results of the synthesis of output blocks (λ function)

			Lambda													
Benchmarks			Without MD							With MD						
Name	In	Out	ALM	ALUT7	ALUT6	ALUT5	ALUT4	ALUT ⇐ 3	Levels	ALM	ALUT7	ALUT6	ALUT5	ALUT4	ALUT ⇐ 3	Levels
beecount	6	4	4	0	4	0	0	0	1	4	0	4	0	0	0	1
dk14	6	5	5	0	5	0	0	0	1	4	0	3	0	1	1	1
dk15	5	5	3	0	0	5	0	0	1	3	0	0	5	0	0	1
dk16	7	3	5	0	2	2	1	2	2	3	1	0	3	1	0	2
dk17	10	3	7	0	1	6	4	1	2	4	0	3	1	0	0	2
dk27	4	2	1	0	0	0	1	1	1	1	0	0	0	1	1	1
dk512	5	3	2	0	0	2	0	1	1	2	0	0	2	0	1	1
donfile	7	1	1	0	0	0	0	1	1	1	0	0	0	0	1	1
ex2	7	2	3	0	2	0	0	1	2	2	0	1	0	2	0	2
ex3	6	2	2	0	2	0	0	0	1	2	0	2	0	0	0	1
ex4	10	9	7	0	5	4	0	0	1	7	0	5	4	0	0	1
ex5	6	2	1	0	1	0	0	0	1	1	0	1	0	0	0	1
ex7	6	2	1	0	1	0	0	0	1	1	0	1	0	0	0	1
lion	4	1	1	0	0	0	1	0	1	1	0	0	0	1	0	1
lion9	6	1	1	0	1	0	0	0	1	1	0	1	0	0	0	1
mc	5	5	2	0	0	1	0	3	1	2	0	0	1	0	3	1
s8	7	1	1	0	0	0	2	0	2	1	0	0	0	2	0	2
s27	7	1	2	0	1	0	0	1	2	2	0	1	0	0	1	2
shiftreg	4	1	1	0	0	0	0	1	1	1	0	0	0	0	1	1

(continued)

Table 15.12 (continued)

			Lambda													
Benchmarks			Without MD							With MD						
Name	In	Out	ALM	ALUT7	ALUT6	ALUT5	ALUT4	ALUT $\Leftarrow$ 3	Levels	ALM	ALUT7	ALUT6	ALUT5	ALUT4	ALUT $\Leftarrow$ 3	Levels
tav	6	4	4	0	4	0	0	0	1	2	0	0	4	0	0	1
train4	4	1	1	0	0	0	1	0	1	1	0	0	0	1	0	1
train11	6	1	1	0	1	0	0	0	1	1	0	1	0	0	0	1
Sum:			56	0	30	20	10	12	27	47	1	23	20	9	9	27

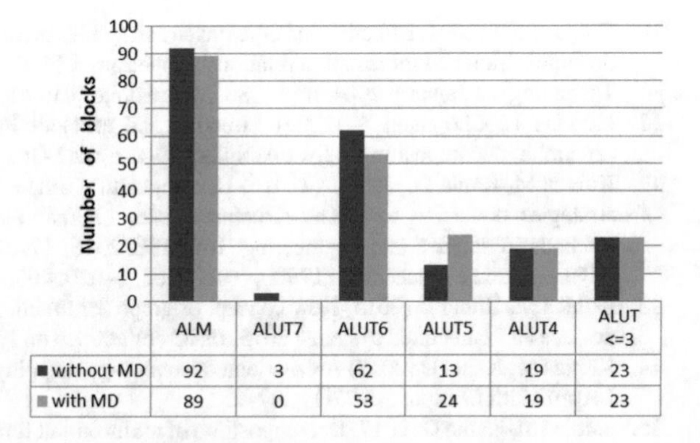

	ALM	ALUT7	ALUT6	ALUT5	ALUT4	ALUT <=3
without MD	92	3	62	13	19	23
with MD	89	0	53	24	19	23

Fig. 15.9 Number of logic blocks obtained after the synthesis of the δ blocks

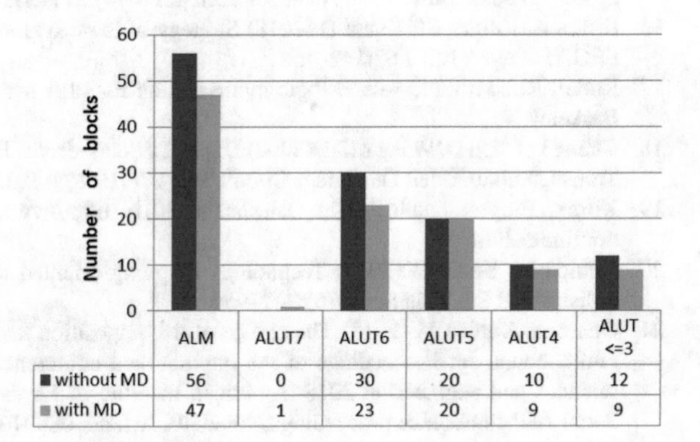

	ALM	ALUT7	ALUT6	ALUT5	ALUT4	ALUT <=3
without MD	56	0	30	20	10	12
with MD	47	1	23	20	9	9

Fig. 15.10 Number of logic blocks obtained after the synthesis of the λ blocks

References

1. Collaborative Benchmarking And Experimental Algorythmics Laboratory, A Benchmark set, 2008. http://www.cbl.ncsu.edu:16080/benchmarks/LGSynth93/testcase/
2. Xilinx Spartan-3 Generation FPGA User Guide (UG331), 2011
3. Opara A, Kubica M, Kania D (2019) Methods of improving time efficiency of decomposition dedicated at FPGA structures and using BDD in the process of cyber-physical synthesis. IEEE Access 7:20619–20631. https://doi.org/10.1109/access.2019.2898230
4. Kubica M, Opara A, Kania D (2017) Logic synthesis for FPGAs based on cutting of BDD. Microproces Microsyst 52:173–187
5. Czajkowski S, Brown SD (2008) Functionally linear decomposition and synthesis of logic circuits for FPGAs. IEEE Trans Comput-Aided Des Integr Circuit Syst 27(12):2236–2249
6. Chen D, Cong J (2004) DAOmap: a depth-optimal area optimization mapping algorithm for FPGA designs. IEEE/ACM international conference on Computer aided design, 2004. ICCAD-2004, pp 752–759
7. Cheng L, Chen D, Wong M (2007) Ddbdd: Delay-driven bdd synthesis for fpgas. In: Design automation conference, 2007. DAC '07. 44th ACM/IEEE, pp 910–915
8. Berkeley Logic Synthesis Group: ABC: A System for Sequential Synthesis and Verification, December 2005. Available: http://www.eecs.berkeley.edu/~alanmi/abc
9. Vemuri N, Kalla P, Tessier R (2002) BDD-based logic synthesis for LUT-based FPGAs. ACM Trans Design Autom Electron Syst 7(4):501–525

10. Chojnacki A (2004) Effective and efficient circuit synthesis for LUT FPGAs based on functional decomposition and information relationship measures, Ph.D. Thesis, Eindhoven University of Technology, 1 January 2004, pp 1–286. https://doi.org/10.6100/ir582596
11. Jozwiak L, Chojnacki A (2003) Effective and efficient FPGA synthesis through general functional decomposition. J Syst Architect 49(4–6):247–265. ISSN 1383-7621
12. Kubica M, Kania D, Opara A (2016) Decomposition time effectiveness for various synthesis strategies dedicated to FPGA structures. 12th international conference of computational methods in science and engineering, ICCMSE 2016, 17–20 March 2016, Athens, Greece, AIP conference proceedings 1790, pp 0300005_1–0300005_4
13. Kubica M, Kania D (2015) New concept of graph for function decomposition, Programmable devices and embedded systems, 2015. IFAC conference on PDES 2015, pp 61–66
14. Kubica M, Kania D (2017) Area-oriented technology mapping for LUT-based logic blocks. Int J Appl Math Comput Sci 27(1):207–222
15. Kubica M, Kania D (2017) Decomposition of multi-output functions oriented to configurability of logic blocks. Bull Polish Acad Sci Tech Sci 65(3):317–331
16. Opara A, Kubica M, Kania D (2018) Strategy of logic synthesis using MTBDD dedicated to FPGA. Integr VLSI J 62:142–158
17. Rudell RL. Multiple-valued logic minimization for PLA synthesis No. UCB/ERL M86–65, Berkeley
18. Cheng L, Chen D, Wong MDF (2008) DDBDD: delay-driven BDD synthesis for FPGAs. IEEE Trans Comput-Aided Des Integr Circuit Syst 27(7):1203–1213
19. Xilinx, Programmable Logic Datasheets, 2016. http://www.xilinx.com/support/documentation/index.htm
20. Kubica M, Kania D (2019) Technology mapping oriented to adaptive logic modules. Bull Polish Acad Sci Tech Sci 67(5):947–956
21. Opara A, Kubica M (2018) The choice of decomposition taking non-disjoint decomposition into account. In: Proceedings of the international conference of computational methods in sciences and engineering 2018. American Institute of Physics, Thessaloniki, 14 Mar 2018, Seria: AIP conference proceedings; vol 2040, Art. no. 080010

Chapter 16
Summary

The motivation to write this book was to show the readers the issues related to logic synthesis for FPGAs. This book is a review of the research carried out by the authors in recent years. The authors presented the results of their work in a partial form in their scientific publications, to which they refer repeatedly in this book. The compilation of these studies in the form of a book allows a broader view of the problem of logic synthesis, and thus it allows more complete conclusions to be drawn from the obtained results, which is the main purpose of this chapter.

The specificity of the FPGA architecture presented in Chap. 1 means that the technology mapping process of a set of logic functions can be oriented towards various goals. Typically, these goals can be the following: limiting the number of logic resources used in an FPGA device (including the configurable logic blocks) or reducing the number of logical levels (which affects the implementation speed). Naturally, the reader can imagine other goals such as minimizing the power consumption or synthesis time. The ideas described in this book are primarily focused on minimizing the number of logic blocks used. The authors refer to both elementary logic blocks, which are LUT blocks in FPGAs; and to complex logic blocks on the example of ALM blocks.

To ensure effective technology mapping, the issue of the representation of logic functions becomes crucial, as shown in Chap. 2. This book focuses on the representation of logic functions in the form of BDD. Underlying this choice is the low computational complexity of the operation, the low computer memory usage for the function representation, and the availability of BDD libraries. The characteristics of using BDD to represent logic functions are presented in Chap. 3.

The problem directly related to the essence of technology mapping is the circuit division problem. The mathematical model of this division is the decomposition of functions. Chapter 4 presents the theories of decomposition and its various resulting models. Chapter 5 reflects these considerations as functional representations in the form of BDD. Generally, a decomposition is associated with the horizontal cut of a BDD diagram. In addition to the classic approach (using a single cutting line), the authors present an approach in which they use several cutting lines simultaneously

M. Kubica et al., *Technology Mapping for LUT-Based FPGA*, Lecture Notes in Electrical Engineering 713, https://doi.org/10.1007/978-3-030-60488-2_16

to implement select decomposition models. This led to the development of a BDD diagram type called the SMTBDD, whose properties were discussed in detail in Chap. 4.

The essence of effective decomposition is the appropriate division of the variables creating bound and free sets, respectively. This leads to the need to periodically search for the best variable breakdown. In the case of BDD, this problem boils down to searching for the best order of the variables in the diagram, as shown in Chap. 6.

It turns out that it is possible to optimize decomposition through the implementation of nondisjoint decomposition. This optimization leads to a reduction in the number of blocks needed to perform bound functions. The essence of this method is to search for variables that can perform the role of these functions. This issue in terms of BDD is discussed in Chap. 7.

Since, as a rule, combination circuits are described by multioutput functions and not any single function, the key is to introduce methods for implementing multioutput functions. Their effective implementation (i.e., one that ensures the effective sharing of logic resources between the individual functions in a team) requires the introduction of effective methods for combining BDD diagrams, which has led to the proposition of a new BDD diagram form, the so-called PMTBDD. Searching for equivalence classes is also key. Chapters 8 and 9 are devoted to the topic of functional decomposition.

To perform effective decomposition, it is necessary to select the appropriate BDD diagram cutting lines. This problem is directly related to the technology mapping problem. The selection of cutting levels is related to the number of logic block inputs. As shown in Chap. 10, it can be somewhat configurable. There are therefore a limited number of possibilities for cutting the BDD diagram, which is why the development of technology mapping assessment methods becomes crucial. The authors introduce a technology mapping coefficient that can be presented in graphic form as a triangle table. The ideas behind using triangle tables in relation to LUT or ALM blocks can be found in Chaps. 11 and 12, respectively.

The developed technology mapping methods for logic functions can also be used for the synthesis of the combined part of the FSM, as shown in Chap. 14.

The results of the work presented in this book are two logic synthesis algorithms focused on the efficient use of LUT-based FPGA resources, which are presented in Chap. 15. The first, dekBDD, uses methods associated with a single cutting line. The second, MultiDec, uses multiple cutting lines. These algorithms have been implemented using the appropriate tools to conduct experiments.

A number of experiments were carried out in which the developed algorithms (dekBDD and MultiDec) were compared with each other and compared with other academic (using different functional description forms) and commercial tools. Experiments were conducted describing the division method using various triangle tables, the impact of nondisjoint decomposition was determined, and the time efficiency of the solutions shown was estimated. The descriptions of ready-made combination circuits and parts of combinational and sequential circuits have been synthesized.

Based on the conducted experiments, the following conclusions can be drawn:

- the results obtained in terms of the number of logic blocks are competitive with both academic and commercial tools,
- both algorithms lead to poorly optimized solutions in terms of the number of logic levels,
- using functional representations in the form of BDD for the decomposition carried out with a single cutting line is of key importance with respect to the synthesis,
- the multiple cut method is much worse than the single cut method of the BDD diagram in terms of the synthesis time,
- the method of estimating the effectiveness of technological mapping is of key importance in the search process for the best technology mapping,
- nondisjoint decomposition is an important element of circuit optimization in terms of the number of blocks,
- taking into account the specific properties of specific FPGA architectures leads to a significant improvement in the efficiency of using logic resources, and
- current commercial tools do not fully use the capabilities of programmable devices, and hence it is possible to improve the solution efficiency of commercial tools.

In summary, the algorithms described in the book can provide good solutions in terms of the number of logic blocks. The synthesis time for a single-cut method is competitive with other solutions. The ideas shown are easy to implement and have a universal character - they can be referred to logic blocks with various configuration capabilities.

The methods shown in the book have some drawbacks as well. It turns out that the developed algorithms work very well for functions with a small number of variables. Unfortunately, for "large" functions, the results are obtained after a very long time, or not at all. Thus, it can be concluded that the described methods should be used for the synthesis of local circuit modules, while the whole should be synthesized using other methods, e.g., by using the AIG network. The next problem is the number of logic levels obtained. It becomes crucial to develop methods for relaxing the critical path of the circuit.

The indicated advantages and disadvantages show that the methods presented in the book have great implementation potential, but they require further optimization and integration with other elements of the synthesis process. Therefore, the reader can find that the directions of further work are closely related to the issues presented in this book. First, it is crucial to focus the described methods on other strategies such as minimizing the number of levels or power consumption. Second, it would be natural to implement the described techniques in the field of cyber-physical system synthesis. Third, the evolution of the described algorithms towards multilevel synthesis seems necessary. Of course, the space for the development of the described ideas is very broad and absolutely not limited to the above issues.

Many years of work related to logic synthesis and technology mapping issues have not led to the solution of all problems. Technological development has led to the situation in which we have very complex devices at our disposal, but we do not know how to use them effectively. An unquestionable success would be the transfer of the synthesis process to a very high level of abstraction. This would allow for fast circuit design, but it would require the use of increasingly sophisticated synthesis algorithms. At a time when we want to reduce power consumption, we cannot afford to use logic resources inefficiently. This makes searching for the best synthesis algorithms an increasingly important challenge.

We hope that reading the book will inspire research teams dealing with logic synthesis issues.

Index

A
ABC, 166, 167, 186, 189
Academic tools, 163
ALM, 115, 133, 140, 148, 191

B
Balanced decomposition, 46
Bdd_apply, 99
Bdd_compose, 89
BDD extracts, 57, 73
Bdd_manager, 67
Bdd_or, 86
Bdd_restrict, 89
Bdd_swap, 67
BDS – PGA 2.0, 166, 186
Binary Decision Diagram (BDD), 25
Binary Decision Programs (BDP), 25
Binary decision trees, 22
Boolean function, 15
Bound functions, 40, 54, 59, 71
Bound set, 39

C
Cache, 33
Canonicity, 34
Cluster, 155
Coding states, 147
Collision problem, 32
Column multiplicity, 39, 45, 54, 59, 120
Column pattern, 79, 96
Combining, 82
Combining functions, 82
Common bound block, 78
Common bound functions, 93
Compatibility graph, 96
Complemented, 33
Complex decomposition, 41
Configurable Logic Block (CLB), 115, 133
Consistency graph, 98
Cube, 15
Cut nodes, 54, 59
Cuts of the BDD, 53
Cutting line, 53, 119

D
DAOmap, 166
DDBDD, 166, 187
Decomposition, 7, 39
DekBDD, 153, 159
Design process, 6
Disjoint decomposition, 71

E
Efficiency coefficient δ, 122
Elimination rule, 26
Equivalence class, 96, 98
Equivalent representations, 34
Espresso, 18
Extract, 60, 73

F
Field-Programmable Gate Array (FPGA), 2
Fitting, 5
FLDS, 163
Free set, 39
From input to output, 48
From output to input, 48

M. Kubica et al., *Technology Mapping for LUT-Based FPGA*, Lecture Notes in Electrical Engineering 713, https://doi.org/10.1007/978-3-030-60488-2

FSM, 147
Full sharing, 77
Functions, 82

G
Garbage collection, 33

H
Hash function, 32
Hypercube, 15

I
Implementation of BDD, 31
Implicant, 17
Internal states, 147
IRMA2FPGA, 166, 187
ITE, 29
Iterative decomposition, 42, 55

K
Karnaugh map, 21

L
Logic synthesis, 201
LUT, 2, 115, 119
LUT4/2, 123, 126
LUT5/1, 123, 126

M
Mealy's automaton, 147
Merge rule, 26
Merginging function, 86
Minterm, 17
Mixed decomposition, 46
Moore's automaton, 147
MultiDec, 159, 189
Multioutput functions, 77
Multioutput implicant, 17
Multioutput minterm, 17
Multiple cut, 60
Multiple cutting, 60, 133, 153
Multiple decomposition, 42, 45, 56, 60, 119, 133
Multi-Terminal BDD (MTBDD), 28, 57, 99

N
Negation attribute, 33
Nondisjoint decomposition, 71, 126, 135, 144
NP-hard, 65

O
Ordered Binary Decision Diagram (OBDD), 25
Ordering variables, 65
Order of variables, 28
Output function, 147

P
Parallel decomposition, 46
Partial sharing, 77, 93
Partition matrix, 39
Prime implicant, 17
Programmable System on Chip (pSoC), 8
Pseudo MTBDD (PMTBDD), 86

R
Reduced Ordered Binary Decision Diagram (ROBDD), 25
Results table, 30, 111
Root table, 59, 73, 98, 120, 135

S
Sequential circuits, 147
Serial decomposition, 46
Shannon expansion, 26
Shared BDD (SBDD), 28, 57
Shared multiterminal BDD (SMTBDD), 57, 59, 73, 97
Sharing, 93
Sharing a bound block, 77
Sharing bound functions, 93
Simple serial decomposition, 39, 53, 60
Single cut, 53
Single cutting, 56
Single cutting line, 153
Single-level decomposition, 46
Slice, 2
Splitting MTBDD, 86
Swap, 82
Symbolic implicants, 19
Synthesis time, 187

T
Table of cut nodes, 78
Table of unique, 33

Technology mapping, 119, 140, 201
Technology mapping table, 143
Transition function, 147
Triangle table, 122, 126, 141, 169
Truth table, 20

U
Unicoding, 93, 97
Unique ID, 67
Unique nodes, 31
Unique table, 89

V
Variable partitioning, 123
Verilog, 191

Z
Zero-suppressed BDD (ZBDD), 27

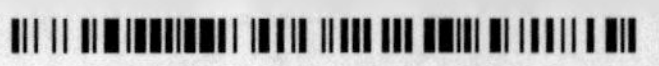

Printed in the United States
by Baker & Taylor Publisher Services